RICK PETERS

FLOORING BASICS

Sterling Publishing Co., Inc.
New York

Acknowledgements

Butterick Media Production Staff

Photography: Christopher Vendetta
Design: Triad Design Group, Ltd.
Illustrations: Triad Design Group, Ltd.
Copy Editor: Barbara McIntosh Webb

Page Layout: David Joinnides
Indexer: Nan Badgett
Project Directors: Nicole Pressly & David Joinnides
President: Art Joinnides

Special thanks to the production staff at Butterick Media for their continuing support. Also, a heartfelt thanks to my constant inspiration: Cheryl, Lynne, Will, and Beth.   R.P.

Every effort has been made to ensure that all the information in this book is accurate. However, due to differing conditions, tools, and individual skill, the publisher cannot be responsible for any injuries, losses, or other damages which may result from the use of information in this book.

Library of Congress Cataloging-in-Publication Data Available

10 9 8 7 6 5 4 3 2 1

Published by Sterling Publishing Company, Inc.
387 Park Avenue South, New York, N.Y. 10016
© 2000 Butterick Company, Inc., Rick Peters
Distributed in Canada by Sterling Publishing
c/o Canadian Manda Group
One Atlantic Avenue, Suite 105, Toronto,
Ontario, Canada M6K 3E7
Distributed in Great Britain and
Europe by Cassell PLC
Wellington House, 125 Strand,
London WC2R 0BB, England
Distributed in Australia by Capricorn Link
(Australia) Pty. Ltd.
P.O. Box 6651, Baulkham Hills, Business Centre,
NSW 2153, Australia

Sterling ISBN 0-8069-5897-9

B
THE BUTTERICK® PUBLISHING COMPANY
161 Avenue of the Americas
New York, New York 10013

Contents

Introduction

In many ways, the flooring in a room can signal its function: ceramic tile in the bathroom, hardwood in the dining room, vinyl in the kitchen, and carpeting in a bedroom, family room, or living room. The flooring in a room can also have a huge impact on the mood or feel of a room. A slate-tile floor in a foyer adds a touch of formal elegance. The natural beauty of a hardwood floor imparts a warm glow in a dining area or family room.

It's too bad that we often neglect flooring as a design element in our homes. We go to great lengths to scrape and paint or wallpaper walls; we spend time, energy, and money on window treatments; yet we ignore the flooring. I think a lot of this might be because most people think that installing flooring is out of their reach—that it requires special tools, costs a lot, and is extremely difficult to do. But the advent of new flooring materials like laminate flooring and resilient sheet vinyl has made new flooring an obtainable goal for the average homeowner.

In this book, I'll start in Chapter 1 by going over the various flooring systems that are available to you. As with any home improvement task, it's a good idea to start with a basic understanding of the underlying structures—in this case the subfloor and floor joists (*page 7*). The types of flooring I'll cover in this book are: carpeting, ceramic tile, hardwood strip flooring, laminate flooring, parquet tiles, resilient sheet flooring, and vinyl tiles (*see pages 8–10 for more on each type*). Before you decide on replacing your existing flooring, check out how to evaluate it first to see whether you may be able to salvage it (*pages 11–12*). I've also included a section on how to estimate materials, including a sample worksheet (*pages 13–15*).

In Chapter 2, I'll describe the general-purpose and specialized tools you need to install flooring (*pages 17–20*). In many cases, you can rent the specialized tools. Then I'll describe the flooring material options; there's even a chart that shows at a glance the differences in average costs and ease of installation (*pages 21–23*).

Chapter 3 takes an in-depth look at one of the most important steps in successfully installing new flooring—preparation. Although it's not always necessary to remove the existing flooring to install a new one, I've included instructions on how to remove common flooring (*pages 25–29*). If the existing underlayment or subfloor needs work, see pages 30 and 31, respectively. The chapter concludes with instructions on how to install plywood underlayment (*pages 32–33*) and cement underlayment or cement board (*pages 34–35*).

The following chapters (4 through 9) provide detailed instructions on how to install the various types of flooring: carpeting (*pages 36–47*), ceramic tile (*pages 48–59*), hardwood flooring (*pages 60–71*), laminate flooring (*pages 72–81*), resilient sheet flooring (*pages 82–93*), and finally, wood and vinyl tiles (*pages 94–103*).

The final chapter, Chapter 10, delves into repairing flooring. A pet peeve of mine, a squeaky floor, can be silenced with any of the variety of techniques shown on pages 105–107. This is followed by details on how to repair most types of flooring: carpeting on pages 108–109, hardwood flooring on pages 110–116 (including sections on how to refinish a floor and remove stains), parquet tiles on page 117, resilient sheet flooring on page 118, and vinyl tiles on pages 120–121. Finally, since ceramic tiles typically require a lot of maintenance, there's information on how to remove stains (*page 122*), replace grout (*page 123*), and replace tiles (*pages 124–125*).

In a nutshell, this book provides all you'll need to know to remove, install, and repair the most common types of flooring. I hope that it helps you with your home improvement adventures.

Rick Peters
Fall 2000

Chapter 1
Flooring Systems

When most of us think of a floor, we envision the top layer—in effect, the decorative covering that in reality is simply a veneer. The "real" floor is hidden under the decorative veneer. It's made up of sturdy plywood or composite panel sub-floor spanning floor joists and may be covered with optional layers of underlayment. Although it's not necessary to delve into span ratings and loads, it is a good idea that you understand the basic anatomy of the "real" floor underneath so that you'll be better able to make wise flooring choices (*see the opposite page*).

After examining the underlying structure, I'll take you through the most common types of flooring used in residential construction: carpeting, ceramic tile, hardwood strip flooring, laminate or "floating" flooring, wood parquet tiles, resilient flooring, and vinyl tiles (*pages 8–10*), everything from where they're commonly used to why you should or should not use a particular floor covering in a room. There's even a chart at the end of the section that quickly compares the advantages, disadvantages, and relative costs of the different flooring options.

Then on to evaluating a floor—how to inspect it and what to look for, along with numerous suggestions for needed repairs, and how to determine when it's time to strip the old flooring off and install a fresh top layer.

If you've decided it's time for a new floor, a worksheet I've included will help you properly estimate materials (*page 13*). The worksheet is a great way for you to compare do-it-yourself costs versus the cost of a flooring contractor to do the job. A great way to determine what's the best way to go.

Finally, there's advice on how to measure a room so that you'll end up ordering the correct amount of flooring for the job. We'll take a look at a relatively simple floor plan (*page 14*) and a complex one (*page 15*). Regardless of the size or complexity, it's important to draw a rough plan of the room so that a flooring contractor can help you determine the correct amount of flooring to order.

Anatomy of a Floor

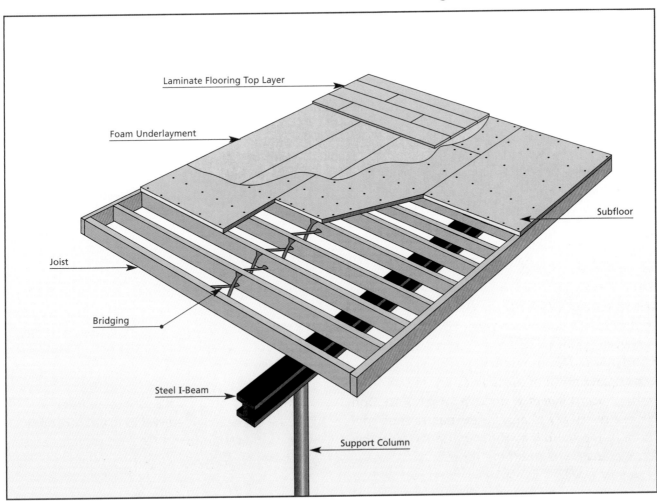

Laminate Flooring Top Layer

Foam Underlayment

Subfloor

Joist

Bridging

Steel I-Beam

Support Column

Regardless of the material chosen as the top layer, the underlying structure of most floors is similar. The most common type in residential construction is the framed floor.

On a ground-level framed floor, the flooring rests on joists that sit on sills along the foundation and is often supported at a midpoint by a steel girder or wood beam.

An elevated framed floor like the one shown *above* is supported by beams that run perpendicular to the joists where the weight of the floor is borne by support columns. In most cases,

the joists are tied together with bridging for extra stability and to keep them from moving from side to side.

Floor joists are covered with some form of subflooring, typically tongue-and-groove plywood, particleboard, or OSB (oriented-strand board). Depending on the type of flooring used, the subfloor may be covered with an additional layer of underlayment, such as cement board (*see page 34*). The top layer of flooring is installed on the underlayment or subfloor and usually rests on some type of cushioning layer such as roofing felt.

Types of Flooring

The type of flooring you choose for a room will depend on appearance, traffic, cost, noise, and safety. The flooring has to fit the room and blend in well with surrounding areas. The amount of traffic the flooring will be subjected to is a factor—high-traffic areas such as kitchens and bathrooms are best served with resilient or ceramic tile. The cost of the flooring will also have a significant impact. Ceramic tile, carpeting, hardwood, and laminate flooring are all moderately expensive. Less-expensive alternatives are resilient sheet flooring and parquet and vinyl tiles. If noise is a concern, softer flooring such as carpeting and cushioned vinyl will deaden sound. Hard, brittle surfaces such as ceramic tile transmit noise but can be muffled with scatter rugs or area rugs. Safety is the final factor that can affect a flooring choice. Many ceramic tiles—especially glazed tiles—can become quite slippery when wet. For an entryway, consider a textured tile instead of a smooth tile to provide better traction.

Carpet Appearance and comfort are the two main reasons many homeowners choose carpeting as a flooring material. It is the most popular flooring choice for living rooms and bedrooms. Conventional carpeting is available in a wide variety of textures and colors (*see page 21*) and requires a pad for cushioning underneath. Cushion-backed carpet, on the other hand, has an integral cushion, built in during manufacture. Cushion-backed carpet is glued directly to a subfloor, while conventional carpeting is stretched and held in place only along the perimeter with special "tackless" strips. Stretching carpet requires special tools and techniques. (*See Chapter 4 for more on this.*)

Ceramic tile Because ceramic tile is extremely durable and is highly resistant to water and stains, it has been used primarily for kitchens and bathrooms. But ceramic tile looks great anywhere in the house. It's very easy to clean and holds up well in high-traffic areas such as entryways and hallways. It is hard and rigid, however, and does conduct noise. Also, ceramic tile is time-consuming to install. You'll have to wait overnight for the mortar that bonds the tile to the floor to set up. And applying grout to fill in the spaces between the tile is messy and also requires overnight drying. (*See Chapter 5 for more on this.*)

Hardwood Even with the new man-made alternatives such as laminate flooring (*see below*), natural hardwood flooring is still a popular choice. Warmth and beauty combined with a durable surface make hardwood flooring the perfect choice for living rooms, dining rooms—almost any room. When properly installed, a hardwood floor can last for generations. When it does start to show signs of aging, it can be refinished (*see page 114*). Installing hardwood flooring requires better-than-average carpentry skills and a few special tools like a flooring nailer and a floor sander. (*See Chapter 6 for more on this.*)

Laminate A newcomer to the flooring market, laminate or "floating" flooring is becoming increasingly popular. It's durable, easy to clean, and easier to install than many of the other flooring choices. What makes it different is that it's not attached to the subfloor. Instead, the edges of the flooring are glued together, and the room-sized glued-up panel "floats" on the subfloor. Typically, a layer of foam is sandwiched between the subfloor and laminate and serves as both a cushion and a noise dampener. Some carpentry skills are required, along with special clamps, to glue the panels together. (*For more on this, see Chapter 7.*)

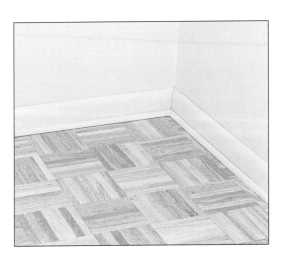

Parquet tile Parquet tile is sort of a hybrid of hardwood flooring and vinyl tiles. It offers the strength and durability of hardwood flooring and the ease of installation of vinyl tiles. Parquet tiles come in a wide variety of patterns and finishes, and all feature tongue-and-groove edges to make installation a snap. Although you can find parquet tiles with self-adhesive backs, you'll achieve longer-lasting results by attaching standard parquet tiles to the subfloor with flooring adhesive. It's messier, but the tiles will stay put. Some carpentry skills are required. (*See Chapter 9 for more on this.*)

Resilient By far the most common type of flooring in use in kitchens and bathrooms today, resilient tile has a well-deserved reputation for durability, resistance to water and staining, and ease of cleaning. Vinyl sheet flooring has a top layer of vinyl bonded to backing, typically felt. This creates a "cushioned" flooring that's durable, comfortable to walk on, and quiet. Resilient flooring is relatively soft compared to many of the other flooring materials and is susceptible to dents and tears caused by sharp objects. Installation requires patience but few specialized tools. (*See chapter 8 for more on this.*)

Vinyl tiles The biggest advantage to vinyl tiles is ease of installation. Twelve-inch squares are a lot easier to apply to a floor than a 12-foot-wide roll. Just like parquet tiles, vinyl tiles are available with self-adhesive backs; but you'll get better results by using flooring adhesive. Vinyl tiles are durable when installed properly and are ideal for high-traffic areas such as kitchens, baths, and entryways. Few specialized tools are required, and the flooring will go down quickly once the starter rows are installed. (*For more on this, see chapter 9.*)

ADVANTAGES AND DISADVANTAGES OF FLOORING TYPES

Type	Uses	Advantages	Disadvantages	Cost
Carpet	bedrooms, family rooms, and hallways	comfortable; available in many colors and textures	shows dirt and wear quickly; requires special installation tools	moderate to high
Ceramic Tile	high-traffic areas like bathrooms and kitchens	extremely durable; easy to clean	time-consuming to install; requires a special underlayment	high
Hardwood	high-traffics areas like dining or living rooms	requires little preparation; goes directly on subfloor	carpentry skills required; special nailing and sanding tools needed	moderate to high
Laminate	high-traffic areas like bathrooms and kitchens	easy to install with average carpentry skills; durable	requires special installation tools, especially clamps	moderate to high
Parquet	dining rooms, kitchens, and living rooms	easy to install; most types are prefinished	not as durable as hardwood flooring; uneven floors cause gaps	moderate
Resilient	most often found in kitchens and bathrooms	easy to install; very durable; cleans easily	shows underlying flaws, so floor preparation is paramount	low to moderate
Vinyl Tile	bathrooms, kitchens, and entryways	very easy to install, especially the self-adhesive type	since there are many seams, not as durable as resilient tile	low to moderate

Evaluating a Floor

Clean, repair, or replace? That's the question most homeowners are faced with when deciding what to do about a tired-looking floor. Quite often, you can breathe new life into the flooring by cleaning or refinishing it. Refresh dull carpeting with steam cleaning. Give hardwood floors a quick makeover by sanding and refinishing (*see page 114*). If the floor is damaged, a little detective work will help you determine whether it can be repaired (*see chapter 10 for more on this*) or whether it needs to be replaced. Evaluating a floor starts with an inspection; depending on the type, you'll be looking for stains, tears, rips, cracks, buckling, or bubbling. Specific concerns for each type of flooring material are listed in the following section. If it's a new look you're after, you'll need to decide whether the old flooring needs to be removed or whether it can stay in place. Some flooring choices require a significant amount of work to be done to prepare for the new flooring—see Chapter 3 for more on floor preparation.

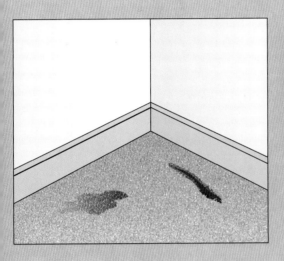

Carpet When evaluating carpet, be on the lookout for stains, tears, and excessive wear. Stains can often be removed with carpet cleaners, or small sections can be replaced with a tuft setter (*see the sidebar on page 109 for more on this*). Tears and holes can often be patched if you've got a remnant handy (*see pages 108–109 for detailed instructions on how to do this*). Excessive wear where the carpet is worn down needs to be replaced. See page 26 for removal instructions and Chapter 4 if you're planning on installing new carpeting.

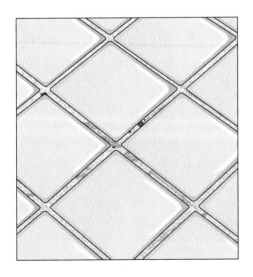

Ceramic tile Ceramic tile, although durable, will eventually start to show signs of wear. The most common trouble area is the grout that fills the spaces between the tiles. Look for loose or crumbling grout and replace with fresh grout (*see page 123*). Mold and mildew are a sure sign that the grout needs to be cleaned and resealed. Inspect for cracked or broken tiles. These can be replaced if desired (*see pages 124–125*). If many tiles are cracked or broken, you may have an underlayment problem; your best bet is to remove the tile flooring, fix the problem, and install a new surface treatment.

Hardwood The Achilles' heel of hardwood flooring is moisture: It can cause the wood to swell and sometimes buckle. As you inspect the surface, check for uneven spots by gently sliding a 4-foot level across the surface. Stains, scratches, and even buckling can be repaired if desired (*see pages 110–116*). If you're planning to install carpet over a hardwood floor, you can usually lay it directly over it as long as it's securely attached to the subfloor. In cases where you're planning on covering it with sheet flooring, it's best to lay down underlayment (*see pages 32–33*).

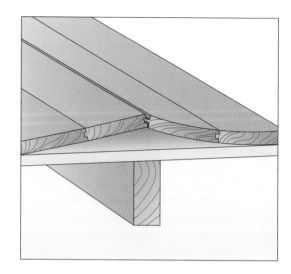

Resilient Over time a number of problems can pop up with resilient flooring that require attention. One of the most common things to look for is air bubbles in the flooring. The best way I've found to find them is to use a trouble light held at a low angle to the flooring—then get down on the floor and look toward the light. Bubbles, tears, and loose sections will be easy to spot. Here again, you can repair these problems if desired (*see pages 118–119*). New sheet flooring can be laid directly over old, as long as the old flooring is smooth and firmly attached to the underlayment.

Vinyl Tile The number one problem with vinyl tiles has to do with all the seams between the tiles. Every seam is an opportunity for dirt and moisture to sneak in and weaken the bond between the flooring and the tile. Consequently, most of the problems related to vinyl tiles, such as curling corners, cracks, and missing sections, occur near or at the seams. Look for these problem spots as you inspect the surface. You can refasten loose pieces or replace entire tiles if desired (*see pages 120–121*). If you're planning on laying down a new surface directly over the old tile, remove loose sections and fill with a leveling compound (*see page 31*).

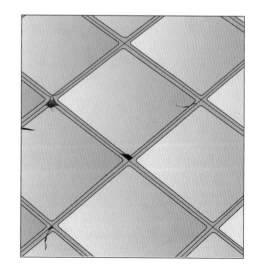

Estimating Materials

Do it yourself, or hire a flooring contractor? The answer depends on your skills and what's more important to you—your time or your money. If you're like most homeowners, you'd rather invest a little sweat equity instead of paying someone else to do the work. One of the best ways to determine how much you can save—and you will save money by doing it yourself—is to get a quote from a flooring contractor and then figure out how much it will cost to do it yourself.

Start by reading through the chapter on the type of flooring you're planning on installing, to get a solid understanding of what's involved. Then use the worksheet below to tally costs. Be as thorough as possible. Sometimes small things likes nails, screws, or adhesive can be surprisingly expensive. If you'll need to rent tools, contact your local rental center for a price list. Be particularly careful when estimating the actual floor covering (*see pages 14–15 for more on this*).

When you're done with the worksheet, compare the cost differential between contractor work and doing it yourself. If the savings is substantial, consider hiring a contractor to handle some of the more labor-intensive or mundane tasks such as removing the old flooring, or sanding and finishing a hardwood floor.

Contractor

Preparation of subfloor
materials $ _____
labor $ _____

Installation
preparation of subfloor
materials $ _____
labor $ _____

Additional charges
moving furniture $ _____
delivery $ _____
finish work (trim) $ _____

Subfloor subtotal $ _____
Installation subtotal $ _____
Additional charges $ _____
Contractor total $ _____

Do-it-yourself

Materials
_____ sheets of underlayment at
$ / sheet = $ _____
building paper or felt $ _____
Finish flooring materials:
_____ square feet at $ ___ /ft.
for a total of $ _____

Tools to buy _____ $ _____
                    _____ $ _____
                    _____ $ _____
Tools to rent _____ $ _____
                    _____ $ _____
                    _____ $ _____

Supplies: adhesive $ _____
              nails/screws $ _____
              sealer/finish $ _____
              baseboards $ _____
              thresholds $ _____

Materials subtotal $ _____
Tools subtotal $ _____
Supplies subtotal $ _____
DIY total $ _____

A Simple Space: a Bedroom

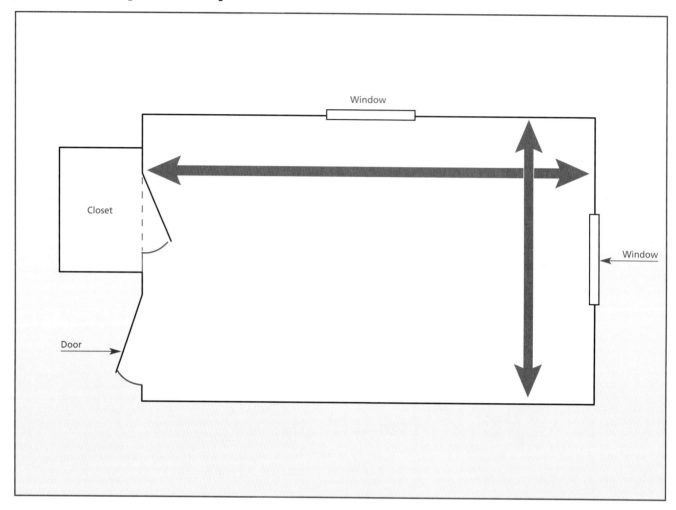

Estimating the floor covering needed for a simple space like the one shown above is fairly straightforward. Start by measuring the overall length and width of the room. Then use graph paper to make a scale drawing of the room. This doesn't have to be an architect's drawing—just a simple representation of the room (include closets, built-ins, and door and window locations).

Next, measure the width and length of any adjoining space that will be covered, like a closet or hallway. Multiply the length times the width for each area, and add them together. For the example shown, the main room is 8 feet wide and 13½ feet long. The closet is 3½ feet wide and 2 feet deep. This is 108 plus 7, for a total of 115 square feet.

You'll always need more than this; it all depends on the type of flooring you're installing—especially if you need to match patterns for a seam. You'll also need extra for overlapping around the edges and so that you can cut the flooring for an exact fit. Ask your flooring supplier for help on figuring how much extra flooring you'll need. Also, if you did get an estimate from a flooring contractor, compare your estimate with theirs and go with the larger number.

A Complex Space: a Kitchen

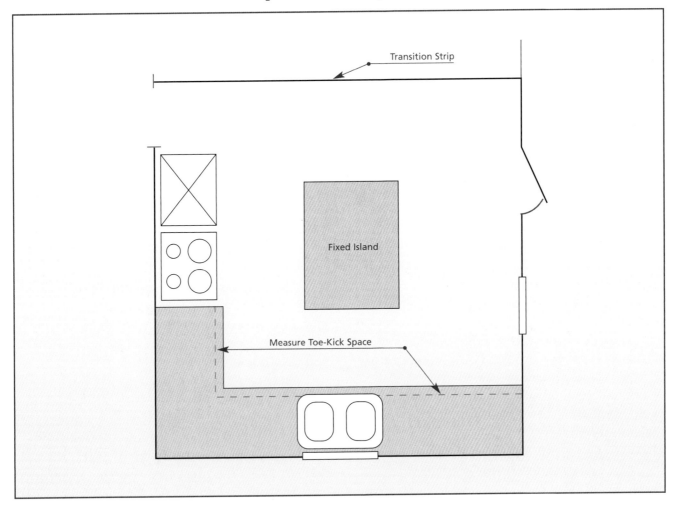

When it comes to estimating the flooring required for a complex space such as the kitchen shown above, you'll need to take into account a couple of things. First, if there are any cabinets in the room, make sure to take your measurements from under the cabinets in the toe space. Second, don't forget to include the area under the appliances if there are any.

In kitchens with islands, it may be simpler for you to remove the hardware that secures the island to the floor and remove the island (with helpers, of course) instead of trying to cut and install flooring around it. If your island doesn't have electrical outlets or plumbing, this may be an option. Regardless of whether you remove it or not, you'll still need to account for the flooring where it rests if you're planning on installing resilient sheet flooring.

If you're installing individual tiles, you can subtract the space under the island from your overall square footage. Here again, check with a flooring supplier as to how much extra flooring you'll need to successfully install it.

Chapter 2
Tools and Materials

The choices you make in both materials and installation tools will have a significant impact on how well the flooring you install will hold up over time. You can purchase the highest-quality flooring available, but if it's not installed properly, it will rapidly degrade. Likewise, low-quality flooring, even if installed properly, won't hold up for long either. In this chapter, I'll take you through the tools you'll need to install each type of flooring and provide advice on what to look for when you're shopping for flooring.

The tools for installing flooring vary greatly from one flooring material to the other. Some materials, like resilient tile, require very few, while others, like ceramic tile and carpeting, need odd-named tools like a grout float or a knee kicker. I'll start by going over the general-purpose tools you'll need for almost any flooring job (see the opposite page), then on to specialized tools for each of the major flooring types: carpet and ceramic tile on page 18, hardwood and laminate flooring on page 19, and resilient, vinyl, and parquet tile on page 20.

The best advice I can give you regarding selecting flooring materials is to go to a flooring supplier or contractor and chat with them about choices and pricing. Many of these folks will let you take home their sample boards overnight to help make your decision easier.

Many home centers now offer small sample squares of flooring—particularly laminate and resilient flooring—and carpeting for less than a dollar apiece. You can usually purchase similarly inexpensive individual vinyl or ceramic tiles. These can help when you know that making a decision will take you a while.

Once you've selected a material, check with your flooring contact about the correct adhesive and/or seam sealers to use. Although you'll likely pay a bit more for the materials from them, you're sure to get quality products that are designed to work together with your flooring material.

General-Purpose Tools

Demolition   Many of the flooring jobs you'll tackle will require some demolition work—that is, tearing out old flooring, underlayment, or a subfloor. Possibly even removing cabinets or a small section of a wall. You'll find the following tools useful for this type of work (*from left to right*): screwdrivers for general dismantling; a pry bar for pulling out stubborn boards and fixtures; a cold chisel or set of inexpensive chisels for chopping out holes in walls or flooring; and a ball-peen or claw hammer.

Measuring/layout   One of the most critical steps in any flooring job is measuring and laying out the grid or starting point. Measuring and layout are also used extensively to ensure that the new flooring is cut to fit properly. The tools shown should be in every homeowner's toolbox: a 3-foot and torpedo level; a 25-foot tape measure; a folding rule for short, accurate measurements; a combination square to check for right angles and for general layout; a chalk line for striking long layout lines; a compass to draw circles; a framing square to check for perfect right angles; and a contour gauge for laying out odd shapes.

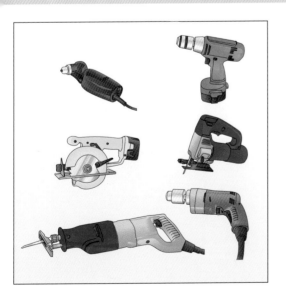

Power tools   Power tools can make quick work of many of the tedious tasks associated with installing flooring. *Clockwise from top left:* a right-angle drill for drilling in tight spots; a cordless drill with a ⅜" chuck for small holes; a saber saw for cutting access holes; an electric drill with a ½" chuck for large-diameter holes; a reciprocating saw for demolition work; and a cordless trim saw for straight-square cuts.

Specialty Tools

Carpet Installing carpet requires quite a few tools. To install the padding, you'll need a staple gun (A) and a sharp utility knife (B). Although you can cut some carpet with a utility knife, there are a couple of specialty tools designed just for this purpose. A row-running knife (C) cuts between the rows of loop-pile carpeting. You may also want to rent an edge trimmer (not shown) for making the final cut around the perimeter of a room.

Stretching and seaming carpet is where you'll come across some odd and very specialized tools—so specialized and expensive that you won't find them at a hardware store or home center. Instead, you'll need to rent them from a rental center. A knee kicker (D) is used to adjust the position of the carpeting on the floor; a power stretcher (E), with its long extensions, allows you to stretch the carpet across a room; and a seaming iron (F) heats hot-melt seam tape so you can join two pieces of carpeting. See Chapter 4 for more on installing carpeting.

Ceramic tile There are three distinct sets of specialized tools that you'll need to install ceramic tile: tile-cutting tools, mortar tools, and grout tools. The type of tile you're installing and the obstacles you'll need to go around will determine what type of cutting tools you'll need. Straight lines are easily cut and snapped on smooth, glazed tiles with a tile cutter (A). A motorized tile cutter (B) allows you to cut thick or rough-textured tiles with ease. You can cut odd-shaped tiles with a rod saw (C) or by removing small bits at a time with a tile nipper (D).

Tiles are affixed to the floor with thin-set mortar. You'll find that a mixing attachment for your electric drill (not shown) will make this task easier. The mortar is applied to the floor with a notched trowel (E), with the notches ranging in size from ⅛" to ⅜". A grout float (F) is used to apply grout, and a sponge (G) wipes off the grout after it's dried to a haze. Finally, a grout saw (H) allows you to remove dried grout for repairs, and a rubber mallet (I) is the best way to set or "bed" a tile in the thin-set mortar. For more on installing ceramic tile, see Chapter 5.

Hardwood strip flooring Installing hardwood flooring requires only a couple of specialized tools: a miter box or power miter saw and a flooring nailer. In addition to these, you'll need some ordinary hand tools: a wood chisel (A) for trimming boards and chopping out sections when doing repairs, a claw hammer (B) for the face-nailing and for installing trim (or you can use an air-powered finish nailer if you have one), a nail set (C) for countersinking the heads of nails below the surface of the boards or trim, and a block plane (D) for trimming strips for an exact fit.

Although you can cut flooring with a miter box and a handsaw (not shown), it'll be a long day (or days, most likely). If you don't own a power miter saw or "chop" saw (E), consider renting one or borrowing one from a friend. Likewise, you can install flooring without renting a flooring nailer (F)—a nifty device that drives nails in at the perfect angle—but it's much faster and easier with one. See Chapter 6 for more on installing hardwood flooring.

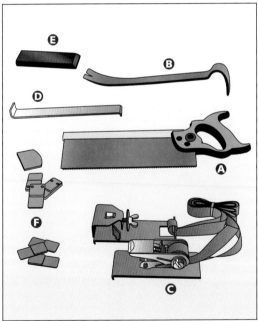

Laminate flooring Many of the tools used for installing hardwood strip flooring are also used to install laminate or "floating" flooring: A miter box and handsaw (A) or a power miter box is used to make straight cuts, a saber saw (*shown on page 17*) is best for cutting curves to get around obstacles, a claw hammer or rubber mallet is for snugging up the joints between boards, and a pry bar (B) is for persuading stubborn boards.

Tools designed specifically for installing laminate flooring are: strap installation clamps (C) for tightening the first couple, or "starter," rows of flooring; a special S-shaped bar for tightening end joints (D); a custom-shaped tapping block (E) that fits over the edge of the flooring so you can drive the boards tight together; and spacers (F) to set the correct gap between the flooring and adjacent walls. Note: Most home centers that sell laminate flooring can also rent you an installation kit. Renting such a kit is definitely the way to go, as a single installation strap clamp costs around $50, and you'll need at least a half-dozen.

Resilient flooring One of the things I like best about resilient flooring is that it hardly requires any specialized tools to install it. As a matter of fact, unless you're planning on a full-adhesive installation, you most likely have all the tools you need on hand in your toolbox. You'll need a staple gun (A) to attach the flooring to the underlayment or subfloor around its perimeter; a putty knife (B) for prying out an errant staple and for general all-around cleaning or scraping off bits of old flooring; a utility knife (C) for cutting tile; a metal straightedge (D) for guiding the utility knife on long cuts; and a small notched knife (E) or trowel for applying adhesive around obstacles, near thresholds, and at the seam, if necessary.

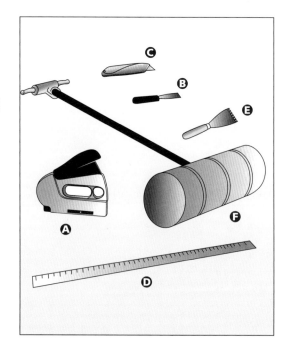

If you are planning on a full-adhesive installation, you'll want to rent a flooring roller (F). A 75- or 100-pound roller rents for less than $20 a day in most areas. A flooring roller presses the flooring firmly into the adhesive to ensure a good bond.

Tile: vinyl and parquet tiles Installation of vinyl tiles basically requires the same tools used to install resilient flooring, shown above. Here again, a flooring roller (A) is a must to ensure a good bond. For repair work, a small scraper (B) comes in handy to remove old flooring, along with a notched trowel (C) to apply new adhesive and a rolling pin (D) or a rubber mallet (E) to bed the new tile firmly in place. Removing tiles that have been glued to the subfloor is hard work. A special floor scraper (F), available at larger home centers, can make the job go a lot quicker. The sharp front edge of the tool is forced under the tile, and a hard push will often pop up a tile or a portion of a tile.

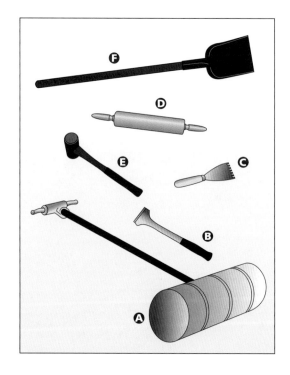

The installation of wood parquet tiles uses some of the tools necessary for installing hardwood strip flooring (*shown on page 19*)—in particular, a miter box and handsaw or a power miter saw, and a saber saw for curved cuts around obstacles. And just as with vinyl tiles, you'll need a flooring roller to bed the tiles in the adhesive.

Materials

Carpet Choosing carpet for your home can be a real challenge. In addition to the wide variety of colors and patterns, it's important to know about the different types available. The four most common types found in homes are: cushion-backed carpet, where the carpet has a foam backing bonded to it; loop-pile carpet, which provides a textured look resulting from the uncut loops of yarn; plush carpet, where the pile is trimmed at a bevel to give it a speckled appearance; and velvet-cut pile carpet, which offers the densest pile of all the carpet types.

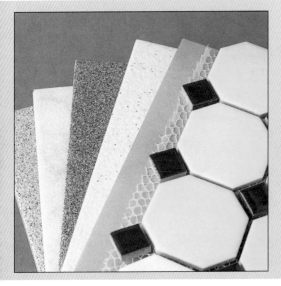

Ceramic tile Ceramic tile is the hardest of all flooring types and is very durable. It's also one of the most expensive and time-consuming to install. Ceramic tiles are all made from molded clay and may be smooth or rough-textured. For flooring, I recommend a textured tile that provides a sound footing. Square floor tiles vary in size from 6" to 12"; other-shaped tiles such as rectangles, hexagons, and octagons are also available. Mosaic tiles typically come in 1" and 2" squares and are sold in 12" square sheets, where smaller individual squares are held together with plastic webbing to make installation easier.

A WORD ABOUT ADHESIVES

Once you've selected a flooring material, it's very important to match the correct adhesive to the flooring. Flooring adhesive is not all the same. As a matter of fact, the warranties on most flooring manufacturers' products will be invalid if you don't use the adhesive they recommend. Read the flooring installation instructions carefully to match the adhesive to the flooring, and make sure to apply it with the specified tool—usually a specific-sized notched trowel.

Also, if you have to join pieces of flooring, it's essential that you use what the manufacturer recommends to splice the pieces together. This is particularly important with resilient sheet vinyl. Not only must you use the recommended adhesive under the seam, but you also need to use the correct seam sealer. Most major resilient flooring manufacturers sell sealer specifically for their product. If you can't find what you need at the local home center, contact a flooring supplier or contractor. If they don't have it in stock, they'll be happy to order it for you.

Hardwood Hardwood strip flooring is available in a variety of woods, widths, and grades. The wood for most flooring is red oak. But you can often purchase other types from a flooring dealer or contractor. Hard maple, beech, birch, ash, sycamore, even pine flooring are some of the more common types. Most strip flooring is ¾" thick and can be bought in 1½", 1¾", 2¼" (the most common), and 2¾" widths. Strip flooring is manufactured with tongues and grooves along its edges and ends. Hardwood flooring grades can get confusing; contact a hardwood flooring contractor for more information.

Laminate When most folks think of laminate flooring, they envision boards or panels that imitate hardwood strip flooring (like the samples shown here). But laminate flooring is available in a wide variety of colors, patterns, and textures. In fact, you can find laminate flooring that mimics almost every other flooring type except carpet. Once you've decided on a style, be wary of inexpensive imitations that look good but won't last long. The primary difference between cheap and quality laminate flooring is the thickness of the top laminate. Quality flooring from a reputable dealer or contractor will have a thick top laminate that will hold up better over the years.

Parquet Wood parquet tiles are made by bonding together strips of wood—typically with glue and a corrugated metal strap embedded in the back of each strip. A variety of patterns are available, but the one shown here is the most common. You can purchase parquet tiles natural or prefinished, in a wide selection of colors. Most parquet tiles are ⁵⁄₁₆" thick, with the edges tongue-and-grooved so they'll slip together easily during installation. Beware of cheap parquet tiles made of low-grade, softer woods. Stick with a hardwood like oak that will stand up better to the abuse that a floor gets.

Resilient Regardless of the color, pattern, or texture, all resilient sheet vinyl is one of two types: full-spread or perimeter-bond. Full-spread flooring has a felt paper backing and is designed for the entire surface to be glued to the subfloor or underlayment with flooring adhesive. Perimeter-bond flooring has a smooth, white PVC (plastic) backing that is secured to the floor only at the perimeter with staples. Because it has no backing, perimeter-bond flooring has some give and take so it can be stretched slightly during installation.

Vinyl Tile Vinyl tiles can be purchased with either self-adhesive or dry backs. Self-adhesive tiles have a wax-paper backing that's peeled off, and the tile is simply pressed onto the floor. Dry-back tiles require adhesive and a flooring roller to achieve a solid bond. Both types of tile are most commonly found in 12" squares. Higher-quality tiles will be thicker than less-expensive versions. Stick with a brand name you can trust, or ask a flooring dealer or contractor for a recommendation.

Ease of Installation and Tools Required

Type	Ease to Install	Tools Required
Carpet	moderate to difficult	staple gun, utility knife, row-running knife, edge trimmer, knee kicker, power stretcher, seaming iron (if necessary)
Ceramic tile	difficult	tile cutter or motorized tile cutter, rod saw or tile nipper, mixing attachment for your electric drill, notched trowel, grout float, sponge, grout saw, rubber mallet
Hardwood	difficult	miter box or power miter saw, flooring nailer, wood chisels, claw hammer, nail set
Laminate	moderate	miter box and handsaw or a power miter box, saber saw, rubber mallet, pry bar, installation clamps, S-shaped bar for tightening end joints, tapping block, spacers
Parquet	easy to moderate	flooring roller, small notched knife or trowel, chalk line, miter box and handsaw or power miter box, saber saw
Resilient	moderate	staple gun, putty knife, utility knife, metal straightedge, small notched knife or trowel (for full-adhesive installations, you'll need a flooring roller)
Vinyl	easy to moderate	flooring roller, small notched knife or trowel, chalk line, utility knife, metal straightedge

Chapter 3
Floor Preparation

Laying new flooring in your home is similar in many ways to painting a room. The actual amount of time you spend painting can often be small in comparison to the time you spend preparing the room for the paint (masking, drop cloths, removing doors, electrical switch plates). This is the same with most new flooring—the time devoted to laying the new covering can be short compared to preparation time (with the exception of ceramic tile, where you often have to wait days between steps).

The amount of time that you'll need to spend preparing your floor for the new covering will depend on what kind of shape the old floor is in and the type of new flooring you're installing. If the old floor covering is firmly attached to the underlayment and/or subfloor, you may be able to lay the new covering directly over it with a minimum of work, such as laying carpeting over an existing hardwood floor. For old flooring that's loose only in a couple of areas, you might be able to get by with just removing the loose sections and leveling the floor. Other situations can require removing the old underlayment,

repairing the subfloor, or adding a new layer of underlayment over the existing floor as when adding cement board in preparation for laying down ceramic tile.

In this chapter, I'll start with how to remove the base molding from a room (and why it may or may not be necessary) and then move on to how to remove each of the different types of flooring: carpet (*page 26*), ceramic tile (*page 27*), resilient flooring (*page 28*), and vinyl or parquet tile (*page 29*). Included in each section are the tools you'll need, along with plenty of tips on how to get the job done as quickly and efficiently as possible.

If more extensive work is needed, there's information on how to remove underlayment (*page 30*), how to level a subfloor (*page 31*), and how to install new plywood underlayment (*pages 32–33*). Finally, there's a section that describes how to lay cement underlayment (also known as cement board)—the best foundation for ceramic tile (*pages 34–35*).

Removing Base Molding

In many cases, the very first thing you'll need to do to prepare a room for new flooring is to remove the base molding from around the perimeter of the room and around any cabinets or built-ins, such as a kitchen island. Not all installations will require this, however. For example, if you're replacing old carpeting with new carpeting, the base moldings (typically baseboards) can remain in place. But on many other installations, such as ceramic tile, resilient sheet flooring, and hardwood flooring, the base molding needs to go.

How you remove molding will depend on the type (wood, tile, or vinyl) and whether or not you're planning on reusing any of it. Regardless of whether you're going to reuse it, you'll want to take steps to prevent damage to your walls. This is easy to accomplish with the judicious placement of a wide-blade putty knife (*see below for both techniques*).

Demolition For situations where you know you won't be reusing the old base molding—it's pretty beat up, or it just won't work with the new flooring—you can remove it quickly with a pry bar. The thing to watch out for is damage to the wall. You can prevent this by slipping a wide-blade putty knife between the molding and the wall. Then you can pry without worry. For cove base molding, loosen a corner with a putty knife and then pull. If the wall is painted, it's a good idea to run the putty knife along the top edge of the molding first to break the paint bond.

Salvage If you're planning on salvaging the base molding, it's best to remove it with two putty knifes, as shown. Slip one against the wall, and insert a stiff-blade putty knife between it and the molding. Now gently pry the molding away from the wall. This takes a bit longer, but it will prevent damage to the molding. On wood baseboards, don't pound the nails out through the face of the molding. Instead, pull them out from behind with a pair of locking pliers (*inset*).

Removing Carpet

1 **Cut around thresholds** Removing old carpeting is fairly straightforward. Start by cutting through the carpet near the door thresholds where the carpet is attached. Depending on the thickness of the carpet and the sharpness of your utility knife, you may be able to cut through both the carpet and the pad at one time. If not, you can always go back after the carpet has been removed and cut and roll the pad. This is also the time to remove any old metal thresholds with a pry bar.

2 **Cut and roll carpet and pad** Next, cut the carpet into strips that will be easy to roll up. I've found that 3-foot strips work best. After cutting the carpet into strips, roll them up and set them aside for disposal. If you cut through both carpet and pad, you may be able to roll them up together. If not, pull up the pad and roll it separately. Most carpet padding is stapled down—just give it a good yank, and it'll pull right up. Note: If you're removing cushion-backed carpet that was glued down, you'll need to remove the adhesive residue with a floor scraper.

3 **Remove tackless strips** If you're planning on laying new carpet, inspect the tackless strips to see whether they can be reused: The points need to be sharp and angled toward the wall. Replace any worn-out sections with fresh strips. If the tackless strips are in bad shape or you're not reinstalling carpet, work around the perimeter of the room, removing the strips with a pry bar and a hammer. Finally, use a putty knife to pry up and lift out any staples remaining from the carpet padding (*inset*).

Removing Ceramic Tile

1 **Break up tiles** Removing ceramic tile is a messy and labor-intensive job. If the tile has been in place for a long time, odds are the adhesive that bonds the tile to the underlayment won't want to release. The best way I've found to remove ceramic tiles is to first break them up into small pieces with a maul or sledgehammer. Make sure to wear eye protection and gloves when you do this. Also, it's a good idea to cover door openings to prevent flying shards of tile from escaping the room.

2 **Scrape away residue** After you've broken the tiles into smaller pieces, use a floor scraper to free them. Hold the scraper at a low angle to the floor to prevent the sharp blade from digging into the underlayment. Stubborn tiles may need to be persuaded with a maul and a wide-blade masonry chisel. If you're not planning on reusing the underlayment, you may be better off cutting through the tile and underlayment with a masonry blade in a circular saw and then prying up sections as shown on page 30.

3 **Use a belt sander if necessary** If you plan to reuse the underlayment, you'll likely have adhesive residue left on it after scraping. A belt sander with a very coarse belt (30- to 60-grit) can be used to grind away the adhesive. Here again, wear eye protection. Since small particles of adhesive, tile, and underlayment will become airborne, you should also wear a dust mask during sanding and when sweeping or vacuuming up the dust.

Removing Resilient Flooring

How easily your old resilient flooring can be removed depends on whether it is a perimeter-installation or a full-adhesive installation (*see page 83 for more on this*). Odds are, if the floor is more than 10 years old, it's a full-adhesive installation. Since perimeter-installed resilient vinyl is attached only along the perimeter and thresholds, it's a snap to remove (*see Step 1 below*). If the flooring was installed prior to 1986, it may contain asbestos; *see page 118 for more on this.*

For flooring that is glued down, consider leaving it in place if possible and simply applying the new covering on top. As long as the old flooring is firmly attached, most new coverings can be applied directly over it. If it's loose in a few spots, you can scrape these off and apply a floor leveler (*see page 31*). However, if the old flooring is loose in many areas, you'll be better off removing it completely (*see Steps 1 and 2 below*).

① Cut into strips and pull For perimeter-installed flooring, use a utility knife to first cut around the perimeter and then cut the flooring into strips about a foot wide. Pull up the flooring by hand, roll it up, and set it aside for disposal. For full-adhesive installations, it's also a good idea to cut the flooring into strips before scraping. You may get lucky and be able to pull up some by hand. If not, you'll at least have created multiple starting points for the blade of the floor scraper.

② Scrape Full-adhesive installations will require a floor scraper and a lot of elbow grease. Keep the floor scraper at a low angle to the floor to prevent it from digging in. In some cases, spraying on a solution of water and dishwashing detergent will help separate the tile from its felt backing. Work in small areas at a time, and take frequent breaks—this work is hard on your lower back. (If you can afford it, this is a great job to hire a flooring contractor to handle.)

Removing Vinyl Tile

1 **Pry and pull** Since vinyl tiles have so many seams and a reputation for not bonding to the subfloor or underlayment very well, they often can be removed fairly easily. As a starting point, look for loose tiles or weak seams that will allow you to slip in a wide-blade putty knife so you can pry up the tile and pull it off. Self-adhesive tiles will occasionally cut up in one piece. Dry-back tiles that were glued down may be more stubborn.

2 **Scrape** Once you've created as many starting points as possible, use a flooring scraper to pry under and lift up the tiles. Keep the angle of the scraper low to the floor to prevent the sharp blade from digging into the underlayment. Wear gloves and goggles to protect your hands and eyes. This work is tough on the lower back, so take frequent breaks, or better yet, get a helper. While one scrapes, the other can pick up the tile residue in an area that's already completed.

3 **Use a heat gun** You can persuade really stubborn tiles to give up their grip by applying a little heat. Use a heat gun to soften the adhesive so you can pry it up with a wide-blade putty knife. Start at a seam and work your way across the tile. Hold the heat gun approximately 4" to 6" away from the tile, keeping it moving from side to side. When you've removed the tile completely, go back while the adhesive is still soft and scrape it off with the putty knife. Wipe this residue off the blade immediately with a clean rag.

Removing Underlayment

1 **Adjust saw** Removing the covering and underlayment at the same time is a quick way to prepare a surface for new flooring. I've found it works best to cut the covering and underlayment into small squares (*Step 2*) and then pry them up (*Step 3*). Adjust the depth of your circular saw to the combined depth of the flooring and the underlayment, as shown. You don't want to cut into the subfloor—just cut deep enough to free the underlayment.

2 **Cut into 3-foot squares** With the circular saw set to the appropriate depth, and wearing safety glasses and gloves, begin cutting the flooring into roughly 3-foot squares. You're not looking for precision here, just tearing up old flooring. When you near the wall, stop cutting and either finish the cut with a reciprocating saw, using a plunge cut, or simply leverage out the uncut piece with a pry bar. If you even suspect that the old flooring contains asbestos, call in a flooring contractor for an assessment.

3 **Pry up** Once the floor covering and underlayment has been cut up into squares, all that's left is to pry it up with a pry bar or crow bar. In either case, jam the blade under the underlayment and pry up. A floor scraper like the one shown on page 20 can also be used. If the underlayment was nailed down, prying should be quick and easy. If screws were used, you'll likely pop the underlayment off and then have to go back and remove them with a screwdriver. When done, go over the entire floor, carefully looking for exposed nails or screws.

Leveling a Subfloor

How flat and level your subfloor is will have a significant impact on how smooth and flat your new floor covering goes down. Regardless of whether you're stripping away old flooring or are planning on laying a new covering on top, it's important to check your subfloor (*Step 1*) and apply a floor leveler as needed (*Step 2*).

Floor levelers are mortarlike cement-based coatings that go down smoothly and set up quickly—some as fast as 10 minutes. Most are ready for the next step in the installation process (such as applying flooring adhesive) in less than an hour.

In addition to leveling out low spots in a subfloor, levelers are a great way to prevent old embossed flooring from telegraphing its pattern over time onto a new flooring such as resilient sheet vinyl. A thin coat of leveler will fill in all the indentations in the old tile to prevent this from happening.

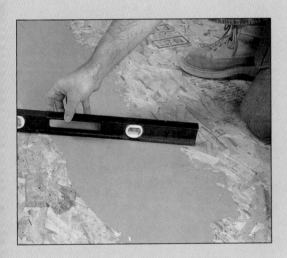

1 Check with a level To determine whether the subfloor is flat, slide a 4-foot level along the surface. A low-angle light (such as a trouble light) placed on the opposite side will help you clearly see highs and lows in the floor. Whenever you detect a problem area, draw a rough outline on the subfloor with a marker to indicate its general shape. High spots are best taken care of by refastening the subfloor to the joists or by sanding off the peaks with a belt sander fitted with a coarse-grit belt.

2 Apply leveling compound Mix the leveling compound according to the manufacturer's directions, and apply it to the floor with a flat-edged trowel. Although most floor levelers dry quickly, you'll want to avoid thick coats. Instead, trowel on multiple thin coats, carefully feathering the edges. When the leveler is dry, scrape off any ridges or high spots with a sharp putty knife and then recheck with the 4-foot level. Repeat as necessary to fill in all depressions.

Installing Plywood Underlayment

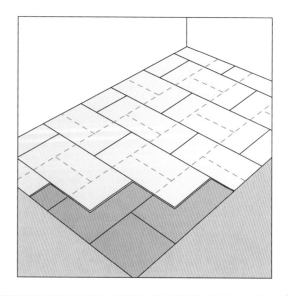

1 **Don't overlap seams** Installing plywood underlayment is a great way to create a flat reference surface for your new floor. It's fairly quick, and there are only a couple things to be aware of. First, make sure the subfloor is level and there are no dips (*see page 31*). Second, to prevent cracks in the new flooring, it's important that the seams of the underlayment don't match up with the seams in the subfloor. Cut the underlayment as necessary to prevent this from happening (*see the drawing at right*).

2 **Start in corner** To install new underlayment, begin by fastening sheets in one corner and work your way across the room. Cut sheets as necessary to prevent overlapping the seams in the subfloor. Fasten the underlayment to the subfloor every 6" or so along the edges and at around 8" to 12" intervals throughout the sheet. Use 1"-long screws or ring-shank nails. Screws have less of a tendency to "pop" over time and are the best choice.

3 **Countersink screws or nails** One of the most important steps in installing new underlayment is to make sure that the nails or screws that you're using to fasten it down are countersunk well below the surface. Any fastener that stands proud of, or is even flush with, the surface can tear new flooring or create an uneven surface. Even if it's not noticeable right away, it will likely cause problems in the future. Take the time to check every fastener, and correct any that aren't countersunk properly.

4 **Cut to fit** Regardless of the room, you'll discover that you need to cut the underlayment to fit around obstacles. Straight cuts are best made with a circular saw. Use a saber saw or coping saw to make circular or curved cuts. Try to cut the underlayment as close as possible to the obstacles to provide maximum support to the new flooring (*see the sidebar below for an easy way to do this*). Instead of cutting around door casings, undercut them as shown on page 62.

5 **Leveling and sanding** To complete the underlayment, mix up some patching compound (or floor leveler) and spread it over seams and screw heads with a wide-blade putty knife or drywall knife. After the compound is dry, scrape off the excess with a sharp putty knife, and sand it flush with the surface with a belt sander and a coarse-grit belt (30- to 60-grit) or with sandpaper and a sanding block. Carefully vacuum up sanding dust before applying the new floor covering.

USING A CONTOUR GAUGE

A contour gauge allows you to accurately copy the irregular shape of an obstacle such as a pipe or door casing and then transfer it onto your flooring. To use the gauge, hold it flat against the floor and press it into the obstacle. The individual fingers of the gauge will conform to the obstacle and will hold that shape via friction. Now you can pull it away and copy the shape onto the flooring with a pencil or marking pen.

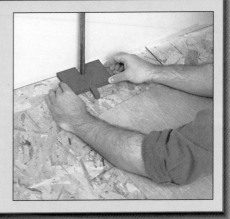

Installing Cement Underlayment

If you're planning on installing ceramic tile flooring in an area where moisture will be a concern, such as a bathroom or a kitchen, you'll need to lay down cement underlayment. Cement underlayment or cement board is rigid and totally stable—even when wet. That's the good news. The not-so-good news is that cement underlayment is expensive, heavy, and difficult to work with. But it is by far the absolute best choice as an underlayment for ceramic tile.

Cement board typically comes in 32" by 60" sheets and is ½" thick. Although you can cut it with a utility knife, a power saw fitted with a masonry or carbide blade is your best bet (*see Steps 3 and 4 on page 35*).

Note: If you're laying ceramic tile in areas where moisture won't be a factor, consider using fiber/cement board (a thin, high-density underlayment) or even plywood. Both of these are less expensive and are easier to work with than cement board.

1 **Apply mortar** Mix your thin-set mortar according to the manufacturer's instructions. You'll find that a mixing attachment for a portable drill will speed up this task. Starting along the longest wall, begin applying the mortar with a notched trowel. In most cases, you'll want to use a ¼" trowel and spread the mortar in a figure-eight pattern. Spread only enough mortar for one sheet at a time. Just as with any other underlayment, make sure that the seams of the cement board don't match up with the seams in the subfloor.

2 **Fasten sheets with screws** Place a sheet of cement board on the mortar with the smooth side facing up. Then secure it to the subfloor with 1½" galvanized deck screws every 6" or so along the edges and about every 8" throughout the rest of the panel. Drive the screws in so they're slightly countersunk beneath the surface of the cement board. Continue like this, working along the wall; then start the next row, making sure the end seams are offset.

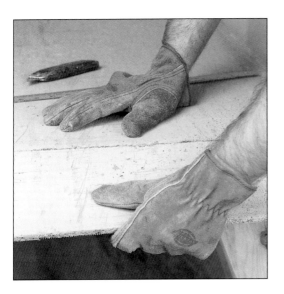

3 **Straight cuts** You can make straight cuts on cement board similarly to cutting drywall. Start by marking the cut with a straightedge and marker or chalk line. Then score a line with a sharp utility knife (it will dull quickly while cutting this abrasive stuff). Next, slide the cement board on your work surface so that the scored line is at the edge, and press down to snap the board (or slip a 2×4 or dowel under the score line and snap it). Finish the cut with the utility knife. Alternatively, use a masonry blade in a power saw.

4 **Obstruction cuts** Quite often you'll need to cut cement board to fit around an odd-shaped obstacle. This is a situation where a saber saw fitted with a masonry or carbide blade is the best tool for the job. If you don't have a masonry blade, you can make short cuts in this stuff with a standard blade, but the cement board is so abrasive, it'll quickly dull the teeth. A tile saw or rod saw can also be used in a pinch to cut this.

5 **Tape and level seams** When all the cement board is down, your final task is to cover the seams and screw heads. To do this, apply thin-set mortar to the screw heads and then place fiberglass mesh tape over the seams. Spread a layer of thin-set mortar over the tape with a wide-blade putty knife or drywall knife. Feather the edges away from the seams to create a smooth surface. Allow the mortar to cure for a minimum of two days before beginning to lay tile.

Chapter 4
Carpeting

Regardless of what type of carpeting you choose, it's all made by pushing loops of yarn up through a backing. The manufacturer can either cut the loops or leave them intact. Although most of the carpet that's manufactured nowadays is made with synthetic fibers such as nylon or polyester, you can still occasionally find carpet made from natural fiber, like wool.

When you're shopping for carpet, the density of the top surface of the carpet, or pile, is generally a good indicator of carpet quality: The greater the number of pile fibers packed into a given area, the better. Denser pile provides a better cushion and holds up better to wear and tear and repeated cleanings.

Regardless of the type of carpet, it's important to recognize up front that installing carpet is a challenging project. Special tools are required, and carpet for the most part is a relatively uncooperative material to work with. For your first carpeting project, I strongly suggest that you stick with a fairly small square or rectangular room. Leave the larger rooms for later, after you have some experience, or call in a professional.

In this chapter, I'll start by going over the two types of carpet you'll find in most homes: conventional carpet, which is stretched with special tools (like a knee kicker and a power stretcher) and held in place with "tackless" strips on the floor around the perimeter of the room, and cushion-backed carpet, which is glued to the floor with adhesive (*see the opposite page*). Then I'll jump into how to install conventional carpet, first looking at installing tackless strips (*page 38*) and installing carpet padding (*page 39*). Then on to the challenges of cutting and seaming carpet (*pages 40–42*). And finally, on to the daunting task of stretching carpet (*pages 43–45*).

Finally, I'll show you ways to handle the transitions from carpet to other types of flooring, such as resilient and hardwood flooring (*page 46*), and complete the chapter with directions on how to install cushion-backed carpet (*page 47*).

Carpet Types

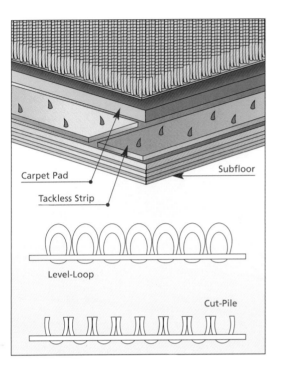

Carpet Pad

Tackless Strip

Subfloor

Level-Loop

Cut-Pile

Conventional Conventional carpet is secured to the floor with tackless strips that are installed around the perimeter of the room. The carpet is "stretched" and then pressed into the pins of the tackless strips to hold it in place. Stretching the carpet requires specialized tools that are best rented. Unlike cushion-backed carpet (*see below*), where the cushion is built-in, conventional carpet lacks padding, which must be installed separately.

There are many different types of conventional carpet available; level-loop and cut-pile are two of the most common. On level-loop carpet (*left, at center*), all the loops of yarn that push up through the backing are the same level. Other variations include carpet where the loops are uneven to create a textured look, either randomly or in a pattern. Cut-pile carpeting has the tops of the loops trimmed off, either straight (*left, at bottom*) or at an angle—depending on the manufacturer's desired effect.

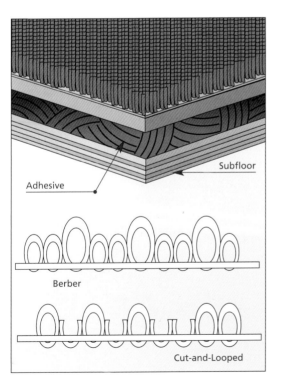

Adhesive

Subfloor

Berber

Cut-and-Looped

Cushion-backed Cushion-backed carpet has a foam backing glued to the back of the carpet foundation. When you install it, you're basically laying the carpet and padding at the same time. This, coupled with the fact that you don't need carpet stretchers to install it (it's glued down), makes it sound too good to be true. Well, it is. Here's the problem: Since cushion-backed carpet is glued down with flooring adhesive, it's a nightmare to remove and replace once it's worn out. There's a lot of scraping involved to remove the old adhesive. Conversely, conventional carpeting can be removed and replaced with comparative ease.

Two of the most common cushion-backed carpet types are Berber (*left, at center*), where the yarn loops' heights vary, and cut-and-looped (*left, at bottom*), where some of the loops are cut. Both types create surface texture, either random or in a pattern. Cushion-backed generally costs less than conventional carpet but is usually lower-quality.

Installing Tackless Strips

1 **Start in a corner** Tackless strips come in a standard width of ¾" and can be used to install carpet on most subfloors. Wider strips are available for installations on concrete. Wearing gloves (the pins on these strips are sharp) and with the angled pins of the strip pointing toward the wall, begin by nailing a strip to the floor in one corner. Maintain a consistent gap between the wall and strip with scrap spacers—check the installation instructions or contact a flooring contractor for the correct gap.

2 **Work around obstacles** Continue nailing tackless strips to the floor around the perimeter of the room. When you encounter an obstacle, such as a radiator, threshold, built-in, or door molding, cut the strips as needed (*see Step 3*) into short lengths so you can work around the obstacle. Try to maintain the same gap that you used for the previous strips, and nail the cut strips to the floor.

3 **Cutting tackless strips** Since tackless strips have angled pins protruding from one face and nails for securing the strips to the floor protruding from the other, they can be a challenge to cut. You can't really lay them flat on a surface to cut them with a handsaw (and if you do, they'll scratch the surface), and they're too thick to cut with a utility knife. The best way I've found to handle this is to cut them with metal snips. Make sure to wear leather gloves whenever handling tackless strips.

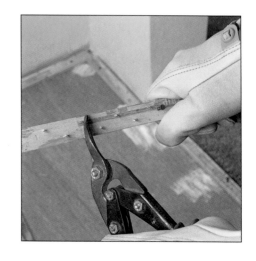

Installing Padding

1 **Roll out** Carpet padding often has a slick side and a rougher side; always install it with the slick side up to make it easier to slide the carpet around during installation. If your padding is wide enough to cover the entire room, roll it out to cover the floor. For narrow padding, start by rolling out a strip to span the length of the room. Cut it to rough length and roll out the next strip. Try not to pull the padding excessively. If it catches, lift it off, don't pull—padding has very little resistance to tearing.

2 **Cut and trim** Position the padding into one corner, and check to make sure it butts up against or overlaps the tackless strips along the length of the padding. Use a sharp utility knife to cut away any excess padding at the corner and along the edges. In most cases, you can cut this freehand; if you're looking for extra precision, use a metal straightedge to guide the knife. The padding should butt solidly up against the tackless strips but not overlap them.

3 **Staple and tape** If you're working with a single piece of padding, work around the perimeter of the room, stapling the padding to the floor every 8" to 10" and at 1-foot intervals throughout the interior of the padding. For narrow pieces, staple the seams to the floor and then run a strip of duct tape to join the pieces together at the seam. Then staple the padding in place as you would for a one-piece padding. Don't go overboard with the staples—the carpet will hold the padding in place for the most part. Staples only help to prevent the pad from shifting or bunching up over time.

Cutting and Seaming Carpet

Cutting and seaming carpet has a fairly well deserved reputation for being a bit nerve-racking. There are a couple reasons for this. First, quality carpet is expensive, and a miscut can be a costly mistake. Second, proper seaming technique requires the use of a hot-glue seaming iron. Since the iron melts the glue only long enough for you to join the pieces together, you must work quickly and without hesitation.

If you do need to join pieces of carpet together, I heartily recommend practicing using the seaming iron to join a couple pieces of carpet remnants together until you get the hang of it. It's not that difficult—it just takes a little getting used to. Also, whenever possible, position seams so they end up in low-traffic areas.

Finally, if you're installing a one-piece carpet in a small room, you may find it easier to unroll the carpet in a larger room and then loosely fold it in half and move it into the room.

1 **Position carpet** With the padding and tackless strips in place, you can install the carpet. In most cases, a roll of carpet is both heavy and cumbersome. Enlist the aid of a helper to bring it into the room and help you unroll it. Start by positioning the roll along one wall, leaving about 6" of excess to run up the wall. Then carefully roll the carpet out until you hit the opposite wall.

2 **Leave 6" excess and cut** When the roll butts up against the opposite wall, mark the back of the carpet at each edge. Then measure up about 6" and make another mark. Pull the carpet away from the wall and snap a chalk line to connect the marks. Cut along this line, using a sharp utility knife or a carpet knife. Make sure to insert a scrap of plywood or other protective material under the folded-back carpet before cutting. Note: If you're installing loop-pile carpet, cut from the top side with a row-running knife (*see page 18*).

3 **Relieve buckling in corners** With the carpet cut to rough length, the next step is to slide it over so the long edge extends up the wall. To do this, straddle the carpet and pull it until it runs up the wall about 6". You will notice immediately that the carpet doesn't want to cooperate at the corners. Use a carpet knife or utility knife to relieve the buckling by slitting the carpet in the corners. Cut just the couple of inches necessary to get the carpet to nestle into the corner; don't cut too far—you'll fit these corners later after stretching the carpet.

4 **Prepare carpet for seam** If you need to join pieces of carpet, start by positioning each piece as you did in Steps 1 through 3. Be sure to include an extra 6" at each wall and 3 extra inches for the seam. Position the cut pieces, making sure the pile all runs in the same direction. Roll back each side of the carpet one piece at a time, and snap a chalk line about 2" back from the factory edge. Then use a straightedge and a utility or carpet knife to cut along the line with a scrap of plywood underneath.

5 **Apply seam glue** One of the tricks I learned from a flooring contractor is to apply a bead of seam glue along the cut edge of the carpet pieces to be seamed. When applied to the edges of the backing, it prevents the carpet from fraying. Seam glue is available at some home centers and most flooring suppliers. It's well worth the effort of trying to track some down if you're installing cut-pile carpet, which tends to "shed" when cut.

6 **Cut the seam tape** The separate pieces of carpeting are now ready to be joined. Although you can try gluing the carpet to the floor at the seams, a better method is to use hot-glue seam tape. You can find this wherever carpet is sold. The glue on the tape is melted with a special seaming iron, available at most rental stores. This type of iron takes a while to heat up, so plug it in and turn it on as you cut the seam tape. Measure and cut a piece of tape to length for each seam.

7 **Use sealing iron** Insert the seam tape under the seam with the adhesive side facing up. Then slide a protective scrap piece of plywood under the tape. Place the iron under the seam directly onto the tape and plywood. In about 20 to 30 seconds, the glue will melt and you can slowly slide the iron along the tape. As you move the iron, let the carpet pieces fall onto the hot glue. Work slowly in roughly 12" sections. Don't get nervous here—the iron won't burn the carpet or the glue, though both will get warm to the touch.

8 **Press edges together** Every foot or so, stop sliding the seaming iron and press the edges of the carpet pieces down into the seam tape with your hands while squeezing the cut edges together. Hold the edges together for 10 to 15 seconds, and move down the seam a few inches. Continue like this until you're close to the seaming iron. Note: When you're done with the seam, use the ridged metal scraper that came with the rental iron to scrape off glue from the bottom of the iron.

Stretching Carpet

Carpet is drawn taut with two specialized tools—a knee kicker and a power stretcher—before it's secured to the tackless strips on the floor. Both are heavy-duty professional-quality tools and are very expensive. Fortunately, both can be picked up at a local rental center for around $100 for a full day. The power stretcher is the main tool that's used to stretch the carpet across the length and width of the room. The less-powerful but more agile knee kicker works best for securing the carpet in corners and around obstacles.

Just as important as the tools is the sequence that you use to stretch the carpet. The sequence shown below gives you the general idea. Short arrows designate the knee kicker, while the longer arrows and the arrows that go from wall to wall call for the power stretcher. Between the weight of the power stretcher and the stiffness of the carpet, it's a good idea to enlist the aid of at least one helper for this demanding job.

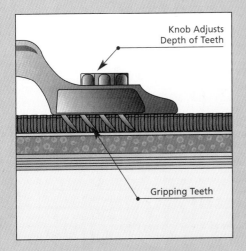

Knob Adjusts
Depth of Teeth

Gripping Teeth

1 Stretchers Although different in size, the business end of both a knee kicker and a power stretcher are similar. Both have two sets of teeth to grip the carpet. One set is comprised of small, fine teeth, set to a given height, to grip the pile of the carpet. The real gripping force, however, comes from a larger set of teeth that can be adjusted via a knob on top of the head. Set the teeth so they'll extend far enough into the carpet to grab its foundation without penetrating through into the carpet padding.

STRETCHING SEQUENCE

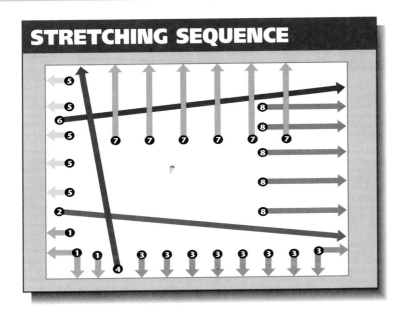

2 **Attach at threshold** Start stretching the carpet by attaching it at the door threshold (if there's more than one threshold, start with the one that will receive the most traffic). Use a knee kicker (*see Step 3*) to push the carpet into position, and then press down on the carpet until the pins of the tackless strips grip the carpet. Most thresholds are near a corner, so push the carpet into place there as well with the knee kicker, and lock it in place on the tackless strips.

3 **Knee kicker** Using a knee kicker takes some getting used to (it's a good idea to practice a bit first before doing the actual installation). Start by positioning the knee kicker about 3" to 5" away from the wall. This way you won't accidentally break one of the tackless strips loose. Holding the knee kicker securely with one hand, strike the pad on the end sharply with your knee to stretch the carpet toward the wall. You may find it'll take a couple of whacks to fully stretch the carpet. Once it's in place, press the carpet firmly down onto the pins of the tackless strips with the head of a hammer.

4 **Power stretcher** To use a power stretcher, begin by placing the power head approximately 4" from the starting wall. Then attach the appropriate combination of extension tubes to the head to reach the opposite wall. At the opposite wall, attach the tail block where the carpet is already secured. Some tail blocks have wheels to make it easier to slide the stretcher along the wall. The wheels also lift the block far enough off the floor so that it fits over most base shoe molding. How much you need to stretch the carpet depends on the type and whether you're stretching across its width or along its length. Ask for guidelines where you purchased your carpet.

5 **Reposition stretcher** After you've stretched a portion of the carpet and pressed it in place on the tackless strips, lift the lever on the power head to release the tension. If you follow the suggested stretching sequence shown on page 43, you'll find that you'll constantly be adding or removing extension tubes as you work around the room. This can be tedious, but it's necessary to get the carpet properly stretched. Power stretchers are heavy, and a helper can make a huge difference as you move and reconfigure the stretcher.

6 **Trim with edger** In a pinch, you can use a carpet knife or sharp utility knife (*as shown*) to trim surplus carpet from the edges of the walls. The best tool for the job, however, is a carpet-edge trimmer. You can rent one of these from your local rental store, or some home centers and most flooring suppliers sell them. To use a carpet-edge trimmer, hold it flat on the carpet and press it firmly into the wall as you move it slowly along the wall. Since it can't quite make it into corners, you'll need to complete the cut with a utility or carpet knife.

7 **Tuck edges** Once all the carpet edges have been cut, work your way around the room, tucking the edges of the carpet into the gap between the tackless strips and the wall. You can do this with a wide-blade putty knife or a special kind of chisel called a stair tool. Depending on the stiffness of the carpet, you may or may not need to use a hammer on the knife or chisel to persuade the carpet edge to cooperate.

Carpet Transitions

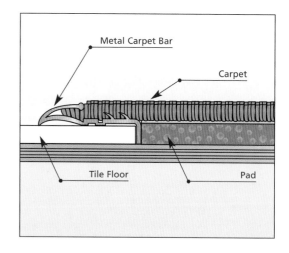

Metal bars Metal carpet bars are used for transitions between carpet and adjacent floor surfaces that are the same height or lower than the carpet, such as vinyl, ceramic tile, or laminate flooring. The carpet bar is first secured to the subfloor with nails or screws, and then the edge of the carpet is inserted under the flange, where it is held in place with angled prongs. Carpet bars are available in standard door-width lengths and can easily be custom-cut with a hacksaw to fit most any threshold.

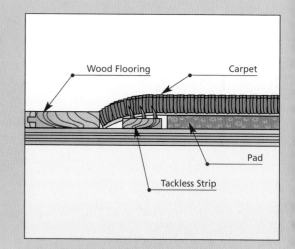

Tackless strip When the adjacent flooring is higher than the carpet bottom (such as when carpeting meets hardwood flooring), use tackless strips to make the transition. All you have to do is treat the higher flooring as if it were a wall, and install tackless strips accordingly. Use a knee kicker to stretch the carpet toward the strip and press it onto the pins of the strip. Then tuck the edge of the carpet into the gap between the flooring and the tackless strip.

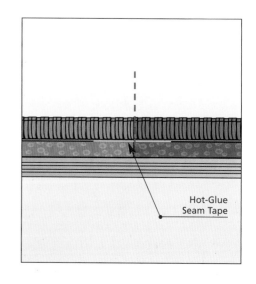

Hot-glue seam tape In situations where you'll be joining the similar-height carpet of one room to that of another at a doorway, join the carpet sections together with hot-glue seam tape (*see pages 41–42 for step-by-step instructions on how to do this*). For cases where the carpet to be joined is of different heights, you're better off installing a wood threshold between the two and then using tackless strips to hold the carpet snug up against the wood threshold as if it were a wall.

Cushion-Backed Carpet

No stretchers required. That's the beauty of installing cushion-backed carpet. The downside is that the carpet is glued down. This is messy and also requires a smooth, flat underlayment. The mess comes from the flooring adhesive that you'll trowel onto the existing floor. Make sure you've got some adhesive remover on hand: Odds are, even if you're extremely careful, some adhesive will surely end up on the carpet.

Also, since cushion-backed carpet tends to be relatively thin, any irregularities in the subfloor or underlayment will telegraph through to the carpeting. In most cases, you'll be better off by laying down a new layer of ¼" plywood underlayment. Another disadvantage to cushion-backed carpet is it's really a pain to remove when it gets worn out. Basically, it has to get ripped or cut into strips and pulled off, and its adhesive scraped off with a floor scraper—hard work.

1 **Roll out and trim** To install cushion-backed carpet, start by rolling out the carpet to cover the floor. For rooms where you'll need to join pieces together, lay the pieces down so the pile is running in the same direction. Snap a chalk line on the floor where the seam will be, and align the edge of one piece with the line. Then place the second piece so it overlaps the edge of the first by ¼". Now you can rough-cut the carpet to fit the room, leaving a 3" margin at the perimeter.

2 **Peel back and apply adhesive** Starting at the seam, pull back both carpet pieces a foot or two. Lay down a thin coat of adhesive with a notched trowel. Carefully fold one piece back and press it in place, working out any air bubbles with your hands. Then unfold the second piece so its edge butts firmly up against the adhered piece. Press it in place, working away from the seam. Now you can fold back one half of the carpeting to the seam, apply adhesive, and press it in place. Cut the edges to fit, and repeat for the remaining piece.

Chapter 5
Ceramic Tile

I've always felt that ceramic floor tile has had an undeserved bad reputation for being difficult to install. It's really not. Sure, it's a bit more complicated than other flooring materials, and it certainly takes longer—mainly because you have to wait for materials to set up, like the mortar that holds the tile to the floor and the grout that fills the spaces between the tiles. But if you break each of the tasks down into simple steps, there's nothing that's really all that difficult.

As a matter of fact, I enjoy installing ceramic tile. It's sort of like a combination of making a mosaic and playing with mud pies during childhood. It's messy, but very rewarding.

In this chapter, I'll start by showing you how to establish the best possible tile pattern for a room, including ways to eliminate narrow border tiles (*see the opposite page*). Then I'll take you step-by-step through the tile installation process—starting with snapping reference lines

and mixing mortar, all the way through to laying and setting tiles (*pages 50–53*).

Then I'll go over the tools and techniques you'll need to cut partial tiles—both to fit along the perimeter and to wrap around obstacles (such as pipes and conduit. I'll start with marking the tiles (*page 54*) and end with different ways to cut them, including using a tile cutter, tile nippers, and a rod saw.

Once the floor tile is cut and laid, you'll need to apply grout to fill in the gaps. The directions on pages 56–57 will take you through this process one step at a time. I've included numerous tips to make this easy but messy job go smoothly.

Finally, there's a section on finishing touches (*pages 58–59*). I'll share techniques you can use that will guarantee that the final eye-catching details will look good as well as ensuring that your tile floor enjoys a long life.

Determining the Pattern

1 **Dry run** After you've chosen the tile for your floor, you'll need to establish what pattern you're going to use. The best way to visualize this is to lay tiles on the floor. Try a square pattern, or possibly one where the tiles run diagonally to the corners (*as shown*). Remember to leave a space between the tiles roughly equivalent to the size of the grout joint you've chosen. Better yet, insert tile spacers (*see page 52*) between the tiles as you lay them.

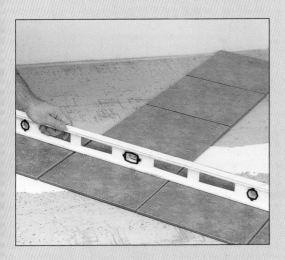

2 **Check for level** When you've decided on a pattern, take the time to check to make sure that the floor is level. Set a 3- or 4-foot-long level on the tiles or subfloor and check for level at numerous places around the room. A floor that's moderately out of level won't present a serious problem unless you're planning on extending the tile up the wall. If this is the case, you'll need to either have the floor leveled or switch to a different type of wall covering.

PREVENTING NARROW BORDER TILES

A common floor-tiling mistake is to ignore the border tiles that are cut to fit around the perimeter of the room. To prevent narrow tiles, draw your tile pattern on a piece of graph paper to scale (*right*). Then on tracing paper, draw the outline of the room to scale. Now place the tracing paper over the tile pattern and move it around to produce the fewest narrow tiles (*far right*). Note how much you'll need to offset the reference lines that you'll use to install the tiles from true center (*see page 50*).

Laying Ceramic Tile

1 **Draw centerlines** Paramount to every successful tile installation is the laying out of the reference grid. Once you've established your pattern (*see page 49*), begin by measuring and snapping your first reference line. Then measure and mark where the roughly centered line that runs perpendicular to the first line goes. Use a framing square to roughly establish this line perpendicular to the first, and snap a chalk line. Then check for true square by using a 3-4-5 triangle (*see the sidebar below*). Adjust one of the lines as necessary until they're perpendicular.

2 **Install battens** Although not absolutely necessary, I like to install battens as a solid starting point for each quadrant. These are nothing more than straight 1×2s that you align with the reference marks that you made in Step 1 and then screw to the subfloor—one for each side. If this is your first time laying floor tile, I heartily recommend using battens. They not only help ensure alignment but also help prevent the tiles from sliding around in the thin-set mortar when you go to "set" them (*see Step 6 on page 52*).

USING A 3-4-5 TRIANGLE

One of the oldest and most reliable ways to check to make sure reference lines are exactly perpendicular is to use a 3-4-5 triangle. To do this, start by measuring and marking a point 3 feet from the centerpoint where the lines cross (make this mark on either line). Then measure and mark 4 feet from the centerpoint on the adjacent line. Now measure the distance from the 3-foot mark to the 4-foot mark. If the lines are truly perpendicular, the distance will measure exactly 5 feet. If it doesn't, the lines aren't perpendicular and you'll need to adjust the position of one of the lines.

3 **Dry-test to check pattern** Here's a way to avoid making a common costly and time-consuming mistake, which I picked up from a flooring contractor. Once your reference lines are snapped and your battens are in place (if you're using them), take the time to make a final dry test of the pattern. Place the dry tiles along the reference lines or battens, and check to make sure everything looks good. I made the mistake once of installing the battens on the wrong side of the reference line, but avoided disaster by dry-testing the pattern before laying the tile.

4 **Spread mortar** When you're sure that the tile pattern is correct, mix up enough thin-set mortar to cover the quadrant you're working in (*see the chart below for types of mortar*). Then use a square-notched trowel to spread the mortar all the way up to the battens or reference lines. Most thin-set mortar manufacturers suggest a ¼" notch for tiles 12" or less in length; larger tiles may require a ½" notch. Avoid working the mortar excessively. What you're looking for here is a consistent mortar with no bare spots.

 # Types of Thin-Set Mortar

Type	Applications
Epoxy mortar	A mixture of sand, cement, liquid resins, and hardeners used when the substrate is incompatible with the other adhesive. Costly, but effective.
Latex- and acrylic-mixed mortar	Similar to water-mixed mortar, but has latex or acrylic added to improve adhesion Also known as latex mortar, it works well in both wet and dry installations.
Medium-bed mortar	Can be applied in layers thicker than ¼" and still retain its strength. Works great for handmade tiles or other tiles that have backs that aren't uniformly flat.
Water-mixed mortar	A blend of portland cement, sand, and additives, also known as dry-set mortar. Used for most tile installations.

5 **Lay first tiles along batten** Now you can begin to lay tiles. Start by positioning the first tile in the corner of the quadrant where the reference lines or battens meet. Press down slightly as you lay the tile to force it into the mortar—you'll set the tile in Step 6. For floor tiles that are over 6" in length or width, twist the tile slightly as you set it in place to help spread the mortar evenly underneath the tile. It's best to press mosaic tiles into the mortar using the edge of a grout float.

6 **Set individual tiles** As soon as the tile is in place, you should "set" or "bed" the tile in the mortar. This ensures that the mortar spreads evenly beneath the tile to afford the best grip possible. The best tool for the job is a soft rubber-faced mallet, or a "dead-blow" mallet as shown here. Although these are made of rubber, one can still break a tile if you hit it hard enough. You're not trying to squeeze out the mortar, just spread it. Multiple light strokes are best: one in the center, followed by one rap in each corner. Then move on to the next tile.

7 **Continue laying tiles** Continue laying tiles along both reference lines. Once these are in place, continue laying tile to fill the entire quadrant. To ensure consistent spacing and even grout joints between the tiles, insert cross-shaped plastic tile spacers between each tile; spacers are available in different sizes to create varying-width grout lines. When the corners of four tiles meet, you can lay the spacers flat on the subfloor. As you near a wall or a batten, insert the spacers on end. If you're laying mosaic tile, use a spacer that's equal to the gaps between the tiles on the sheets.

8 **Level the tiles** After you've got an entire quadrant filled with tile, use this contractor's tip to level the tiles. Nail a scrap of carpet around a 3-foot-long piece of 2×4. Then position the "leveler" on three tiles at a time, and give the top a sharp rap with a mallet centered along its length. Slide the leveler slowly across the tiles, working from one edge to the other. Overlap the leveler onto the tiles adjacent to the first set you leveled, and repeat the process. This is the simplest way to prevent the corner of a tile (or tiles) from protruding above the others.

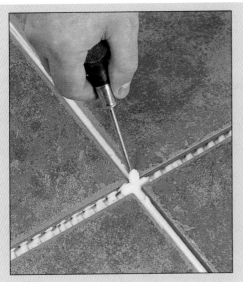

9 **Remove spacers** With all the tiles set firmly into the mortar, you can work around the floor, removing the plastic cross-shaped spacers from between the tiles. Although you could still manage to pry them out of the mortar once it has set up, it's a lot easier to do while it's still soft. Use an awl or a pair of needlenose pliers to remove each of the spacers. Be careful not to nudge the tiles out of position; if you do, carefully return the tile to its proper position.

10 **Laying a saddle** If you're planning on installing a saddle to serve as a transition between different types of flooring, now is the time to do it. Most saddles are made of solid-surface materials and come in standard door-width lengths. If you do have to cut one, use a tungsten-carbide blade in a hacksaw or circular saw. Apply thin-set mortar beneath the tile and set it in place, leaving a gap between the tiles and the threshold equal to the width of one grout joint. Let the mortar cure for at least 24 hours before applying grout (*see page 56*).

Cutting Large Tile

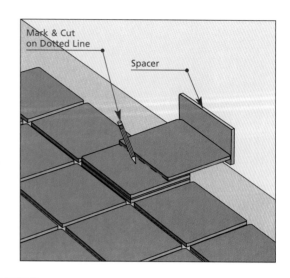

1 **Mark** As your floor tile nears a wall, you'll most likely need to cut it to create a border. Unless cover or trim tiles are to be installed, the first step is to place the tile to be cut directly on top of the full-sized tile nearest the wall. Then place another tile on top of the tile to be cut, and place a spacer the same thickness as the grout lines against the wall. Slide the top tile over until it butts up against the spacer and, using the edge of the top tile as a guide, mark a line on the tile to be cut.

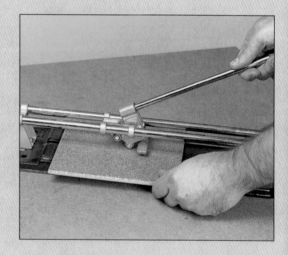

2 **Score** To cut the tile, slip it into a tile cutter so the line you marked on the tile is directly under the scoring wheel and so the edge of the tile butts firmly up against the top of the tile cutter. Now, pressing firmly down on the handle, draw the wheel across the full length of the tile. If the wheel doesn't score a continuous line, reposition the wheel at the top and repeat. Many tile cutters have a built-in breaker bar that can be used to snap the tile. Follow the manufacturer's directions if this is the case on your tile cutter; otherwise, go to Step 3.

3 **Snap** An alternative way to snap a tile is to place the scored line of the tile directly over a scrap of dowel. Then press firmly down on both sides of the tile at the same time to break the tile cleanly. Even if the scored line was true, you'll often find that the tile doesn't break perfectly clean. In situations like these, you can use a tile nipper (*inset*) to nip off the excess ragged tile, leaving a smooth edge. Tile nippers can remove only small portions at a time, so go slow and have patience.

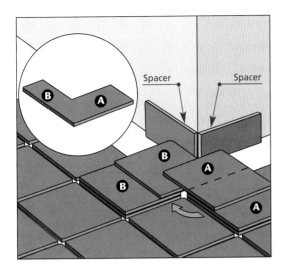

4 **Partial tiles** Partial tiles, for where you need to cut a notch to fit around an outside corner or other obstacle, are a bit trickier than the straight cut shown in Step 1. Fortunately, you can use the same procedure to mark the tile as you did in Step 1. The only difference is that you have to set up and mark the tile on both sides of the corner (*see drawing at left*). Here again, remember to insert a spacer between the wall and the tile equal to the thickness of one grout line.

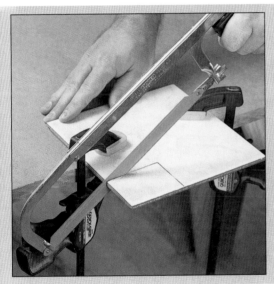

5 **Use a hacksaw** Since the cuts you need to make on a partial tile don't go all the way across a tile, you can't use a tile cutter to score a line and snap the tile. Instead you'll need to use one of a variety of tools that are designed just for this job. If all the cuts you've got to make are straight, the least expensive option is to use your hacksaw fitted with a tungsten-carbide blade, like the one shown here. But this is slow-going and requires a lot of elbow grease. If you've got a lot of tile to cut, consider renting a tile saw.

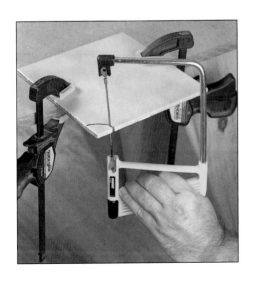

6 **Tile saw** A tile saw or rod saw is designed to make curved cuts in tile. Its blade—basically a rod coated with tungsten-carbide bits—can be moved in any direction as you cut. As you make your cut, you're basically grinding away the tile; full-length, light, even strokes work best. Just like the tungsten-carbide blade in the hacksaw, this work goes slow and requires a lot of effort. To cut a shape out of the center of a tile, first drill a hole with a masonry bit and then thread the rod through the hole and cut.

Grouting

Contrary to popular belief, grout doesn't help hold or lock tiles in position—the thin-set mortar does that. Instead, grout's main purpose in life is simply to fill in the gaps between the tiles. Grouting should only be done once the thin-set mortar has had sufficient time to set up—a bare minimum of one day; two days is safer. Since gravity is working for you and not against you (as when tiling a wall), grouting a floor is easy to do, but messy. You'll need a few specialized tools, like a grout float (*see page 18*) and a set of old work clothes that have seen better days.

One of the biggest mistakes I've seen a homeowner make is to mix up a batch of grout large enough to cover the entire floor, and then discover it has set up halfway through the floor. Mix up only enough grout to work a 4- or 5-foot-square section of tile at a time. This not only allows you to take your time to do the job right, but it also makes mixing easier, since you're working with smaller batches.

1 **Apply grout with a float** Start by mixing up a batch of tile grout according to the manufacturer's instructions to cover roughly a 4-square-foot area. (Many manufacturers suggest letting the grout sit at least 15 minutes before applying it.) Start in one corner and pour a modest pile of grout on the tile. Then, using a grout float, spread the grout over the tiles and into the gaps between them. Press down firmly to force the grout into the joints. Don't worry about neatness here; just fill the joints completely with grout.

2 **Skim off excess** Now use the grout float to skim off the excess grout. Hold the float at an angle so that the bottom edge acts like a squeegee. Skew the float diagonally as you wipe it across the tiles. This way the edge of the tile can safely span the joints without falling in and squeezing out the grout from the joint. Continue working the area until the majority of the grout has been removed. Then go over it one more time with the grout float held nearly vertical to scrape off as much as possible.

3 **Wipe off with a sponge** Although removing the remaining grout from the tile with a sponge is easy work, it's time-consuming. One reason why it takes so long is that you need to rinse out the sponge with clean water frequently. Just as you did with the float, wipe the sponge diagonally over the tiles. Wipe over each grout joint only once; repeated wiping can pull the grout out of the joint. When finished, repeat Steps 1 through 3 for the next area of the floor. After the grout has dried about four to six hours, use a soft cloth to buff away the grout film.

4 **Tool joints with a dowel** Some tile manufacturers suggest that you "tool" the grout joints to give them a uniform appearance. Although you can purchase a special tool specifically designed for this, a short length of dowel will do. The diameter of the dowel must be large enough to span the joint. The larger its diameter, the smaller the concave depression it'll make in the grout. To use the dowel, position it over a grout joint and draw it along the tile with gentle pressure.

5 **Seal** Grout is porous and needs to be sealed to prevent staining and mildew from growing. Following the manufacturer's directions, apply seam sealer to the cured grout. To prevent the sealant from trapping moisture in the grout, most manufacturers suggest waiting two to four weeks before applying their product. Apply the sealer carefully to the grout joints only, using a small sponge or a sash brush. Wipe up any excess sealer immediately with a clean, dry rag.

Finishing Touches

I'm sure that whoever first said "God is in the details" wasn't thinking about ceramic floor tile. But it is the details or finishing touches that can really determine how good or how bad a tile flooring job looks when the installation is complete. Not only that, it's the detail work that can have a huge impact on the overall longevity of a floor—things as simple as sealing the gap between the border floor tiles and the base trim, and sealing the tops of the trim tiles to prevent damaging water from sneaking behind the tile.

Just like the way the trim on your house catches the eye, the trim around a floor's perimeter will be extremely noticeable. That's why it's so important to take your time here. Unfortunately, it's very common to rush through this final task to get the job done. Don't do it. Instead, take a break, a deep breath, whatever, to relax and take the necessary time to complete this critical stage of the job.

Tile baseboard Most flooring tile manufacturers make base trim tiles to match their floor tiles. These typically feature a small cove at the bottom to serve as a smooth transition from wall to floor and to make cleaning easier. But sometimes you may not like how they look, or they simply may not offer trim tiles to match. In cases like this, you can install a row of bull-nose tiles like the ones shown here. You can pick a similar color, or pick a contrasting color as an accent.

Back buttering The easiest way I've found to install trim tiles is to use a technique called "back buttering." Instead of trying to apply thin-set mortar or adhesive to the wall (which is both awkward and difficult to spread at a uniform height), you spread the mortar or adhesive to the back of the tile just as you spread butter or jam on a piece of toast. Keep the adhesive about 1/4" to 1/2" away from the top edge to prevent squeeze-out when you press the tile in place.

Threshold options There are a number of ways to treat the transitions from ceramic tile to other flooring materials at the thresholds. One option is to install a ceramic threshold or saddle (*see page 53*) that's held in place with thin-set mortar. Another possibility is a wooden threshold, like the one shown. Wooden thresholds work well, look good, and are easy to install. Simply cut them to the correct length, and nail them to the subfloor.

Caulk Regardless of the type of base trim you've installed, it's important to seal the gap between the trim and the floor tiles and between the top edge of the trim tiles and the wall. Depending on the type of trim, this may be grout (for base-trim tiles) or caulk (for bull-nose tiles, cove base molding, or wood trim). Sealing the gaps prevents dirt and moisture from working under the tiles and causing damage to the wall or the floor. Choose a caulk that matches the trim, and apply it with a caulk gun. I'd suggest a silicone-based caulk: It's long-lasting and flexible.

Caulk smoother After you've applied the caulk, go back over it with a caulk smoother to even out the bead and remove any excess caulk. You can buy a plastic caulk smoother specifically for this, but I've always found that a moistened fingertip works fine. Make sure to wear protective gloves if you're planning on doing this, and use gentle pressure and move your hand in a steady, even motion. Wipe off any excess with a clean, damp rag or a damp paper towel. Avoid the common mistake of working the caulk too much—one pass works best.

Chapter 6
Hardwood Flooring

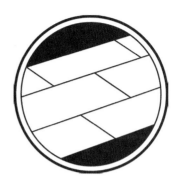

One of the best ways to add warmth and beauty to a room is to install a hardwood floor. There's nothing quite like the natural glow that real wood can add to a room. Hardwood strip floors are rugged and will last for generations if well maintained. Since most strip flooring is ¾" thick, it can be sanded and refinished many times before it needs to be replaced. But all of this beauty and warmth comes at a price. Installing hardwood strip flooring is hard work, and it requires some modest woodworking skills. If you haven't worked with wood much, I'd suggest installing a laminate floor (*see Chapter 7*). Laminate flooring is thinner, is easier to cut, and is glued together to "float" on the subfloor.

If you have some skills with wood and you're up to the challenges of strip flooring, let's take a look at what it entails. But first, an important note about strip flooring. One of the most common mistakes I've seen homeowners make when installing a hardwood floor is not allowing the flooring to thoroughly acclimatize to the room before installing it. Wood is hygroscopic—that is,

it responds to moisture in the air. Wood shrinks and expands to match ambient moisture and seasonal changes in humidity. Since most flooring has a higher moisture content than the interior of your home, it's imperative that you allow it to adjust. In most cases, this takes three to six weeks! That's why I mention it now. Purchase the flooring, set it in the room, and wait. This simple step will help your flooring go down smooth and remain that way over time.

In this chapter, I'll start by going over common ways to fasten strip flooring to the subfloor (*page 61*). Then I'll take you through the steps you'll need to prepare a room for strip flooring, everything from layout lines to trimming doors and casings (*see pages 62–63*). After that, we'll jump into installing flooring—using a flooring nailer, and doing it manually (*see pages 64–68*). I've also included a brief section on how to glue down flooring instead of nailing it in place (*page 69*). And on pages 70–71 are instructions on how to sand and then apply a finish to your new floor.

Installation Options

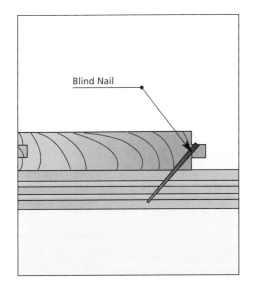

Tongue-and-groove The most common way to install strip hardwood flooring is to nail tongue-and-groove flooring to the subfloor as shown. Each strip has a tongue running the length of the strip on one side and a matched groove on the other. A nail is driven into the tongue at an angle and is covered by the groove in the next strip; this is known as blind-nailing, since the nail is "invisible" in the finished flooring. A less common (and less reliable) installation is to glue the strips in place with flooring adhesive (*see page 69*).

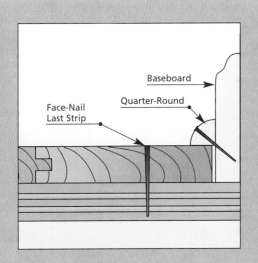

Face-nailing Another less common installation method is face-nailing. Tongue-and-groove strips are still used, but they're nailed directly through the face of the strip to the subfloor. You'll occasionally find face-nailed flooring in an older home where the owners needed the floor to go down quickly, or they couldn't afford the additional labor costs of blind-nailing—blind-nailing by hand takes considerably longer. Nowadays, flooring nailers make it easy to blind-nail, and face-nailing is used only near walls (*see the drawing at left*), where the wall interferes with the nailer (*see page 64*).

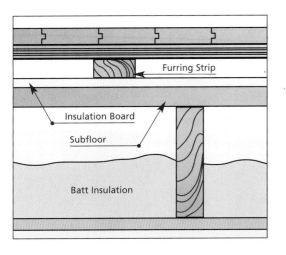

Soundproofing The final installation method used is to "float" the floor on furring strips and insulation board. This method is used when noise is a concern; it helps to muffle footsteps on a floor overhead and also reduces noise from below. Attach ½"-thick insulation board to the subfloor with silicone caulk. Then glue 1×3 furring strips to the insulation board with construction adhesive. Finally, screw a ½" plywood subfloor to the furring strips and nail the hardwood flooring to the plywood. If you're after the ultimate in noise reduction, install insulating batts between the joists in the ceiling below.

Preparing for Flooring

There are a number of things to do to prepare a room for hardwood strip flooring. Since hardwood flooring is the thickest of all types of flooring that you could install, the biggest challenge is dealing with the added thickness. Not only does this include thinking through the threshold transitions to other types of flooring (*see page 67*), but also determining how to handle doors (which won't swing open onto the thicker floor) and door casing. Doors must be trimmed (*Step 1 below*), and casings as well, to accept the new flooring (*Step 2*).

Next, you'll need to lay down building or felt paper to serve as a moisture barrier (*Step 3 on the opposite page*). Then take the time to carefully lay out and mark the joists so that you can nail the flooring to them for the added strength (*Step 4*). Finally, it's best to frame obstacles—such as the hearth of a fireplace, a radiator, or any area you won't be covering—with strips of flooring (*see Step 5*).

1 **Trim doors** Remove any doors that open into the room, and shorten them so they'll work with the added thickness of the new hardwood floor. Add ¼" for clearance to the thickness of the flooring, and cut this off the bottom of the door. For wide doors, use a straight-edge as a guide for a circular saw. If you need to trim a hollow-core door (like the one shown), make sure the bottom perimeter strip is wide enough to handle the trim. If it isn't, you can make the cut, peel off the outer door skin from the strip, and glue the strip back in place.

2 **Trim casing** Instead of trying to sculpt the flooring to fit around the intricate edges of the doorstop and door casings, it's a lot easier to "undercut" the casings and slip the flooring beneath them. To do this, place a scrap of flooring next to the casing to be trimmed. Then with a handsaw lying flat on the scrap as shown, cut into the casing. Be careful not to scratch the wall as you near the end of the cut; one way to help prevent this from happening is to cover the teeth on the end of the saw with a few strips of duct tape.

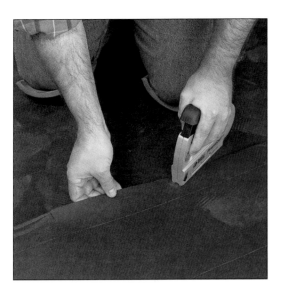

③ Lay felt paper Since wood constantly reacts to changes in humidity—it swells and shrinks as it changes—anything you can do to keep moisture from seeping into the wood will help. First, lay down a layer of felt or building paper on the subfloor before installing the flooring. Start laying felt paper along the longest wall and work your way across the room, overlapping the strips 3" as you go. A couple of staples will prevent the ends from curling and keep the paper in place until the flooring is installed.

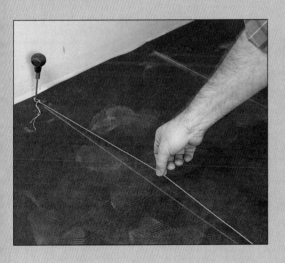

④ Locate and mark joists Once the felt paper is down, locate the joists with a stud finder and mark their locations on the felt paper. Then snap a chalk line at every joist location to serve as nailing guides. Next find the midpoint of the room and snap a line to mark the center. Now measure equal distances from the ends of the center-lines to roughly ½" from the starting wall and snap a line there; this is where you'll start laying strips. Doing this ensures that the highly visible strips in the middle of the room will look straight even if the room is out of square.

⑤ Frame borders It's a good idea to frame special sections of the floor with "border" strips before beginning to lay down the flooring. Border strips are typically mitered to wrap around an obstacle, such as the tile hearth shown here or some other area that won't be covered with flooring. Install the strips so that the groove in the flooring is facing out toward the room. This way you can interlock the tongues of the other flooring strips as they're installed. Border pieces are always face-nailed to the subfloor. Countersink the nail heads, and fill them with putty before applying a finish.

Installing Flooring

Although the basic technique for installing hardwood strip flooring isn't difficult, it does require some ability with wood. And it is hard work. Even once you've got the first couple of rows in place, you'll be faced with a lot of cutting, trimming, and nailing. It's the nailing that gets me. Even with a rented flooring nailer, it's tough on the lower back. But in my mind, the natural beauty of a hardwood strip floor is well worth it.

Before you get started laying the strips, there's one really important thing to do: Check the flooring to make sure its moisture content is low enough to install. You can purchase a moisture meter for around $100 from most woodworking supply stores (to locate one, search the Internet for "moisture meter"). If you've allowed the flooring to acclimatize as I suggested in the introduction to this chapter, it should read between 6% and 9%. If it does, it's ready to be installed. If not, wait a bit longer and test it again.

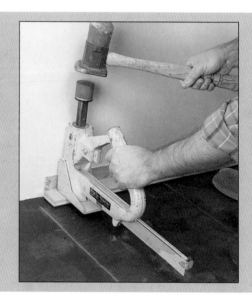

1 **Face-nail starter strip** The first rows of flooring must be face-nailed since the flooring nailer needs about a foot of wall clearance to work. With the groove facing the wall, align a long strip with the starter line you snapped in Step 4 on page 63. Some rental centers supply a special face-nailer (like the one shown here) that's designed to work close to the wall. Following the manufacturer's instructions, nail the starter strip in place. You can also face-nail manually by drilling a slightly smaller hole than the nail and then driving the nail in place. Use 1½" nails, and nail into the joists where possible.

SCRIBING

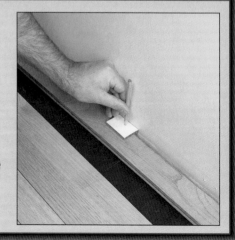

Depending on how square a room is, the first or final piece of flooring may need to be custom-cut to fit. The simplest way to mark the board for this cut is to "scribe" it. There are special tools for this, but a piece of posterboard and a pencil will do.

Here's how. Start by aligning either the tongue or the groove edge with the edge of the nearest strip to the wall. Then measure the largest gap between the wall and the strip and add ⅛". Poke a hole in a piece of posterboard to span the gap. Now butt the posterboard against the wall and run it slowly against the wall as you press down on the pencil. The pencil will scribe the unevenness of the wall onto the strip.

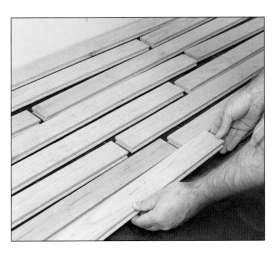

2 **Arrange fields** After you've face-nailed the first course, lay down at least two more courses, but blind-nail these to the joists by hand (*see the sidebar below*). Here again, you're too close to the wall to use the flooring nailer. Make sure to leave a ½" gap wherever the flooring meets a wall. Next, lay out or "rack" seven or eight rows or "fields" of flooring, staggering end joints in adjoining rows by at least 6".

3 **Nail down with flooring nailer** After the third course is down, you should be able to use the flooring nailer. Start by tapping the strip with the rubber end of the mallet to position it and close gaps. Then hook the lip of the nailer over the top edge of the strip and strike the plunger with the metal end of the mallet to drive in a fastener. Drive in fasteners at the joists and halfway between them, and near the ends of the strips. Make sure that the nailer drives the fastener at least ⅛" below the surface. Keep a nail set on hand to countersink any fasteners that protrude.

NAILING WITHOUT A FLOORING NAILER

I heartily recommend using a flooring nailer whenever possible. When you can't— such as when you're too close to a wall, or when you're doing some patchwork—you can nail the strips in place the way it was done before nailers were around: with a hammer. There are a couple of tricks to this. First, predrill an angled pilot hole for the nail through the tongue and into the subfloor as shown. Select a bit that's slightly smaller than the nail's diameter. Predrilling will ensure that the nail goes where you want it and prevents splitting at the same time. Next, drive the nail in until it's slightly above the surface. Then use a nail set to drive it into and slightly below the surface of the tongue.

4 **Drive strips together with scrap** Because wood flooring reacts to changes in moisture, odds are that no single strip that you've purchased will be absolutely straight. This means you'll likely need to persuade most strips into place to close up any gaps. If you've rented a flooring nailer, some mallets that come with the nailer have a rubber face on one end that can be used for this. Don't be tempted to strike the edge of the flooring with a hammer—it'll dent the edge, leaving an unsightly gap. Instead, use a short scrap of flooring, or a "beater" block, and a hammer or mallet to drive the strips together.

5 **For warped boards, apply leverage** No matter how high-quality the flooring is that you've purchased, you'll still come across pieces that are warped or slightly twisted. As long as you can persuade the strip to behave long enough for nailing, you should be able to use it. The simplest way that I've found to muscle a board in place is to drive a chisel or screwdriver into the subfloor at an angle against the edge of the strip to be persuaded. Then lever the chisel or screwdriver into the strip to close any gaps. With the strip in position, have a helper drive a couple of fasteners into the strip.

6 **Pry edges of final boards** When you have worked your way across the floor to the opposite wall, you'll encounter a couple of problems. First, you won't be able to use the flooring nailer because of wall clearance, so you'll have to face-nail the last strip in place. Second, you'll find that the cramped quarters will make it difficult to close any gaps in the last piece. To do this, place a scrap of wood against the wall and insert a pry bar between the scrap and the strip. Lever the prybar to close any gaps, and have a helper face-nail the strip in place.

7 **Cut to fit** In most cases, when you reach the opposite wall, you'll need to trim the strip to fit the gap. If the wall is uneven, use the scribing technique shown on page 64 to mark the strip to match the wall. To cut a strip along its length, clamp it to a pair of sawhorses so the edge to be cut extends out past the sawhorse. Then use a circular saw or trim saw (*as shown*) to cut the piece. Stay on the waste side of the line you marked, and smooth any rough edges with a piece of sandpaper or a block plane.

8 **Reducer strip** Wherever strip flooring meets a threshold, you'll need to make a transition to the flooring in the next room. One way to do this is to use a reducer strip. You can buy these wherever strip flooring is sold, in a variety of thicknesses and styles and with varying notches in the bottom to fit over most flooring. (One edge of the reducer is grooved to lock into your flooring.) To install a reducer strip, cut it to length to match the width of the door opening. Then nail it in place, countersink the nails, and fill the holes with putty before applying a finish.

9 **Expansion strips** If you encounter places where the strip flooring meets ceramic tiles, metal doors, or a laid stone floor, install a strip of cork between the flooring and the nonwood material. The cork fills the gap yet still allows the strip flooring to expand and contract with seasonal changes in humidity. In most cases, you'll want ¾"-thick cork. You can find this at some home centers that sell hardwood flooring, or you can order it from a flooring supplier or contractor.

10 **Reversing direction** If you're planning on extending the strip flooring out of the room and into a hallway or closet, you'll need to reverse the direction of the tongues and grooves in the flooring. This is necessary so that the flooring nailer will be able to get in close to the wall in the closet or hallway. Most flooring dealers sell a special slip tongue that's used to join together flooring groove-to-groove as shown. In a pinch, you can make your own by cutting ¼"-thick strip of hardboard to a width of around ½".

11 **Trim** After you've sanded and applied a finish to the floor (*see pages 70–71*), you can install trim—usually a 3½"-tall baseboard of some sort. Miter or cope the corners of the baseboards, and nail them into the wall studs with 1½"-long finish nails. Countersink the nail heads under the surface, and fill the holes with putty before applying a finish. Depending on the expansion gap you used between the strip flooring and the wall, you may need to cover any remaining gap with a shoe molding (*see Step 12*).

12 **Shoe molding** Most flooring professionals install a molding to cover the expansion gap between the baseboards and the strip flooring—usually ¾" shoe molding (*see the photo at right*). To install shoe molding, it's best to predrill a ¹⁄₁₆" angled pilot hole for the nail. Measure and cut the shoe molding to length, and fasten it to the baseboard with 1½"-long finish nails. Use a nail set to countersink the nails below the surface, and fill the holes with putty before applying a finish.

Gluing Flooring

1 **Apply adhesive** Another way to install strip flooring is to glue it to the subfloor instead of nailing it in place. Gluing it directly to the floor doesn't allow the wood to expand and contract as well as nailing it does, and it's a lot more work to remove. Start by preparing the room as you did for nailing (*see pages 62–63*). Then, working with about 2-foot sections, apply a flooring adhesive to the floor with a V-notched trowel. After you've applied the adhesive, wait approximately 20 to 30 minutes before applying the flooring to allow the glue to become tacky.

2 **Fit and set** Just as you did for nailing, place the first strip about ½" away from the starting wall, with the tongue facing the wall. Tap it with a mallet to set it in the adhesive. Next, apply carpenter's glue sparingly to the tongue end of the next strip, and insert it in the groove in the first strip. Wipe away any glue squeeze-out immediately with a clean, damp cloth. Continue laying strips like this until you near the edge of the adhesive, and repeat Steps 1 and 2 until the floor is covered. Make sure to stagger the ends of adjoining boards by at least 6".

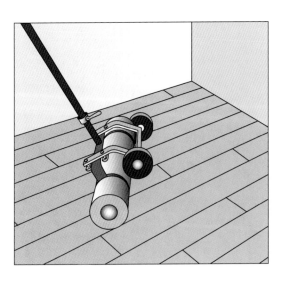

3 **Press with flooring roller** To get the best bond between the strips and the subfloor, you'll need to press the strips firmly into the adhesive. The best tool for this is a rented flooring roller. A word of caution here. As you move the roller across the flooring, it's likely to cause the glue that you applied to the tongues to squeeze out. It's imperative that you wipe this up immediately with a clean, damp cloth to prevent the glue from interfering with the varnish used to seal the floor later. Have on hand a bucket or two of clean water just for this, and wipe the roller off regularly to prevent spreading the glue.

Sanding and Finishing

To complete the flooring job, you'll need to first sand the floor to smooth it out, leveling any high points in the strips and removing any imperfections, and then apply a couple coats of clear finish. As an optional step, you can also apply a stain to the floor to match existing furniture or flooring. Both of these jobs are messy but are easy to do with the right tools.

For sanding, this includes a rented floor sander—either a drum sander for floors that are rough or an orbital sander (like the one shown below) for flooring that's relatively smooth to begin with. Although both sanders will generate dust, the drum sander will create huge clouds of it. The main reason for this is that it's a lot more aggressive than an orbital sander. This is good if you need to remove a lot of wood, but it can also damage your flooring if you leave it in one spot too long (*see Step 1 below*). You'll also need either to rent an edge sander for sanding the perimeter or to use a rented or borrowed random-orbit sander (*see Step 2*).

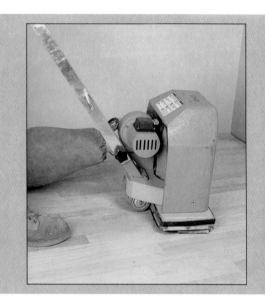

1 **Floor sander** Although an orbital sander takes longer to sand a floor than a drum sander does, I recommend it for first-time users: It's just a lot easier to use. And unlike a drum sander, it won't grind a depression in the floor if you leave it in one place too long. (See page 114 for instructions on how to prepare a room for sanding.) To use an orbital sander, place a sanding sheet or screen under the pad, plug it in, put on a dust mask, and go. Keep the sander moving and about 3" away from the walls (you'll sand the edge areas in Step 2).

2 **Detail sanding** Although you can rent an edging sander to sand along the perimeter of the room, I don't recommend them. Most edging sanders are just large disk sanders. Disk sanders, by their very rotating nature, will leave swirl marks on a wood floor. A better swirl-free alternative is to use a random-orbit sander. These small, portable tools are surprisingly aggressive with a coarse sanding disk or screen but are designed not to leave swirl marks. The one shown here even has a built-in dust filter to help hold down the dust.

3 **Vacuum** The number one cause of a poor finish on a new floor is lack of attention to vacuuming. Removing all the sanding dust is critical for a good finish. Start with a wide nozzle and vacuum the entire floor. Then empty the vacuum canister, clean the filter, and wait an hour or two for the dust you just stirred up vacuuming to settle back onto the floor. Now go over the floor one more time with the vacuum. If you notice a lot of dust in the air when you're done, clean the vacuum again and repeat the process until the air is free of dust and there's nothing to settle on the floor.

4 **First coat** When you're positive the floor is as clean as possible, it's time to apply a finish. Contact your flooring supplier for a finish recommendation. In most cases, they'll suggest a polyurethane designed specifically for floors. I like to use a lamb's-wool applicator to do this. It holds a lot of finish and makes it easy to obtain a smooth, even coat. Start in one corner and work your way across the room, taking care to overlap strokes. When you're done, wrap a plastic grocery bag around the applicator so you can use it later to apply additional coats.

5 **Subsequent coats** After allowing the first coat to dry following the manufacturer's instructions (usually overnight), you can apply the next coat. Depending on the finish you're using and the amount of gloss you're after, you may need to put on three or four coats. At the very minimum, you should apply two coats. Let the final coat cure for a minimum of two days before moving any furniture into the room. To prevent scratches, apply felt pads to furniture feet, especially for heavy items such as sofas and chests.

Chapter 7
Laminate Flooring

I must admit, I was quite skeptical when I first encountered laminate flooring years ago. The concept of a "floating" floor that doesn't get nailed or fastened to the subfloor made me nervous. I wondered what would keep it from buckling or moving around over time. And it's so thin, how could it possibly be durable? But the more I learned and worked with this new material, the less skeptical and more impressed I became.

When installed properly, laminate flooring does not buckle—it lies perfectly flat. And when I discovered that it's made of materials similar to those used to make kitchen countertops (like fiberboard, cellulose paper, and hard melamine resins), I knew it could stand up to a lot of abuse over time. Also, since the planks are glued together, you effectively create one large panel that can swell or shrink as a single unit when humidity changes. Here's where not attaching it to the subfloor is a benefit—the flooring can expand or contract without buckling, as hardwood flooring is prone to.

As with hardwood flooring, it's important to purchase your laminate flooring in advance of the intended installation date so the planks can acclimatize to the room. Unlike hardwood strips, which take three to six weeks for this, unopened cartons of laminate flooring only need to be placed in the room where they'll be installed 72 hours in advance.

I begin this chapter by fully describing what laminate flooring is and how it "floats" on the subfloor (*opposite page*). Then I'll cover the types of underlayment and how to install them (*page 74*). Next, there are step-by-step directions on how to lay the first or "starter" course (*pages 75–77*) and the additional rows (*page 78*). I've also included sections on how to work around obstacles (*page 79*); techniques for finishing up, such as removing dried glue and installing thresholds (*page 80*); and adding shoe molding to conceal the expansion gap necessary for the floor to react to changes in humidity (*page 81*).

How It Works

and check for moisture. If you find beads of moisture on the underside of the plastic, you've got a moisture problem; call in a flooring contractor for advice. If the plastic is dry, you can install laminate flooring; just make sure to first lay down a vapor barrier before installing underlayment.

All floors require some type of underlayment to be installed before the flooring is laid down. Laminate flooring manufacturers offer a variety of options for this. You'll find that cork panels and rolls of foam are the most common. The cork panels offer the best soundproofing and insulating properties but can be expensive. In most cases, the rolled foam will work fine (*see page 74 for more on this*). The subfloor should be level and free from dips and high spots. Any depression greater than 3/16" should be filled with a leveling compound (*see page 31*).

Floating floors One of the biggest benefits laminate flooring has to offer is that it can be laid down over most existing flooring. Not having to first remove the old flooring will save you both time and money. There are a couple of exceptions to this. First, laminate flooring can't be laid over carpeting; that must first be removed. Second, if you're planning on installing laminate flooring directly over concrete, you'll need to do a moisture test. To do this, cut a couple of 2-foot squares of plastic and duct tape them to various areas on the floor. Wait 72 hours

WHAT IS LAMINATE FLOORING?

All laminate flooring is basically made up of the same four parts. Here's what they are from the top down. The top or "wear" layer is cellulose paper that's impregnated with clear melamine resins. Directly below this is the design layer, which can be a photo or other pattern printed on paper and is also impregnated with resins for strength. The middle section or core is most often made from fiberboard—finely ground up wood chips or sawdust that's coated with resins and pressed into sheets under heat and high pressure. The bottom or stability layer, along with the top layer, creates a sandwichlike moisture barrier that protects the core and helps keep it from warping due to changes in humidity.

Wear Layer

Design Layer

Core

Stability Layer

Underlayment

All laminate flooring requires the installation of an underlayment to deaden sound, prevent glue from fastening the planks to the subfloor, and cushion your step. As I mentioned on page 73, you'll need to install a vapor barrier as well if you're installing laminate flooring over concrete. The easiest underlayment to install is foam; there are a number of types available.

Closed-cell polyethylene is the most common and provides fair noise reduction and cushioning. Another version, often referred to as 2-in-1 foam, combines a vapor barrier with a foam cushion. This allows you to install one layer over a concrete floor instead of two separate layers. The 2-in-1 foam is applied to the subfloor with the film side down. If you want the ultimate in soundproofing, insulation, and cushioning, use solid or rubber-backed polyester. It comes in boards or rolls and is more expensive than foam.

1 **Roll out** To install foam underlayment, place the cut end of the roll against the wall in one corner of the room and unroll it. Cut it to length with a sharp utility knife or a pair of scissors. To prevent tearing the underlayment as you work, most manufacturers suggest laying one row of foam at a time and then covering it with flooring. With solid underlayment, cover the entire floor by butting the panels together and taping across the seams every 8" or so. Leave a ¼" gap at the walls and around other obstacles.

2 **Tape seams** Once you've covered a section of foam underlayment with flooring, roll out the next row and cut it to length. Butt the edges of the foam together and use duct tape to join the seams. Make sure that the foam doesn't overlap. Cover this section with flooring, and repeat the process for the remainder of the floor. You may find that a staple in each corner helps keep the foam in place until the flooring goes down. Just make sure to remove the staples before installing the flooring.

Starter Course

The first three rows or "starter course" of laminate flooring you lay down will have a huge impact on the overall success of the installation. It's important that these planks go down flat and straight so that the rest of the planks will be easy to install. Start by choosing the longest wall, a straight wall, as your starting point.

There are a couple of things that are critical for laminate flooring to go down correctly. Since the planks are glued together and not attached to the floor in any way, it's imperative that you use sufficient glue and pressure to create strong, tight joints (*see the sidebar on page 77 for gluing guidelines*). Pressure is best supplied with specialized strap clamps, but in a pinch you can make some simple wedges that will do the job (*see page 78*).

1 Dry-fit Starting in one corner, lay down a plank with the groove facing the wall. Insert spacers between the plank and the wall to create the appropriate expansion gap. Most laminate manufacturers specify a ¼" gap. You can use scraps of ¼"-thick plywood or a combination of plastic spacers that come in the installation kit (*as shown here*). Push the plank and spacers firmly against both walls as tightly as possible. Lay as many full-length planks as the wall permits.

2 Mark the length When you're near the end of the wall, you'll need to cut a full plank down to fit. The easiest way to do this is to place the end plank so its groove butts up against a spacer at the wall. Then use a combination square and a pen to draw a line where the plank meets the ends of the full plank, as shown. Extend this line to the end of the plank and over onto its back.

3 Cut to fit Once you've marked a plank, you can cut it with a circular saw or trim saw. Here's a trick I've used for years. If you've got some scrap ¾"-thick foam insulation board lying around, you can use it to make quick and easy cuts. Just slip a piece of foam board under the plank to be cut. Then set the blade depth to just slightly more than the thickness of the plank. Pressing the plank firmly against the foam board, make your cut. The foam board will prevent chip-out by supporting the area around the blade. After the foam board has been scored a half-dozen times by the blade, flip it and use the other side.

4 Dry-fit next two rows With the first row in place, go ahead and dry-fit the next two rows. Follow the manufacturer's instructions on how to stagger the joints. The most common method used is to cut the first plank in the second row so it's roughly two-thirds the length of a full plank. The first plank in the third row is cut one-third the length of a full plank. This produces a pleasing pattern and ensures that the floor joints are far enough away from each other to create a strong floor. Make sure to cut the correct end of the plank—the end that butts up against the wall.

5 Apply glue Once you've cut and fit the first three rows, you're ready for glue. Carefully slide the planks apart, keeping them in order. Apply glue to one row at a time, starting with the first row. Put glue on the side and edge of each plank, unless the edge butts up against a wall (*see the sidebar on the opposite page for gluing guidelines*). Insert the tongue of one plank into the groove of the adjoining piece. Squeeze the planks together with hand pressure, and set them in place on the floor. Wipe off any excess glue with a clean damp cloth or sponge.

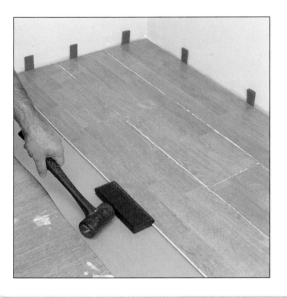

6 **Tap together** After the first row is glued together, begin gluing up the second row. As you slide each piece to mate with the first row, use the tapping block (that came with the installation kit) and a hammer or mallet to force the planks together and tighten up the joints. To do this, slip the profiled edge of the tapping block over the tongue of the plank. Then tap it gently with a mallet; slide the block along the plank, tapping as you go, until any gaps close up. You can use the block on the ends of the planks as well.

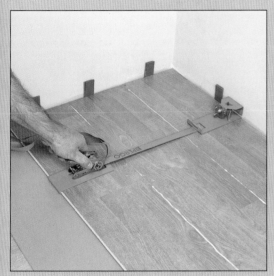

7 **Strap clamps** The next step is to use strap clamps to hold the planks together so the glue can set up (about an hour). You can usually rent an installation kit where you purchased your flooring. It'll contain 6 to 10 strap clamps, a tapping block, and spacers. Since these clamps cost around $50 each, renting is the way to go. Slip one end of the clamp over the plank near the wall. Fit the end with the ratchet lever over the opposite side. Remove any slack from the strap, and ratchet the lever to pull the joints tight. Don't go overboard; tighten them just enough to get uniform glue squeeze-out (*see below*).

HOW MUCH GLUE?

One of the most common problems first-time installers of laminate flooring make is not using enough glue. Most flooring manufacturers sell their own glue—use it. It's designed specifically for the correct absorption into the core of the plank to create the strongest possible joint. Virtually every manufacturer will direct you to fill the groove completely with glue. It's easy to tell whether you've applied the right amount—just observe how much squeezes out as you connect the planks. What you're looking for is a uniform bead of glue along the joint line (*see the drawing*). If you don't achieve this, pull the plank back out and apply more glue. After the planks are joined, wipe off the excess with a clean, damp cloth.

Additional Rows

1 **Use wedges** If you find that you're short a strap clamp or two and need to apply more pressure, here's a simple clamp based on opposing wedges that you can make in minutes. Simply cut some ¾" scraps of plywood or other sheet material into simple wedges. They don't have to be exact—a rough-cut taper will do. Fasten one wedge to the subfloor temporarily with screws so that its tapered edge faces the plank. Then slip another wedge between the fastened wedge and the plank, *as shown.* A few gentle taps with a hammer will close any gaps.

2 **Scribe final boards** When you've worked your way across the room and you are ready to install the final planks, you'll most likely need to trim them to fit the space between the wall and the last full plank. To mark the plank for trimming, set the plank to be cut on the full plank nearest to the wall, making sure to orient the groove so it will mate with the full plank below it. Then use a scrap piece of plank and a pen to scribe a line to match the wall, as shown. Cut the plank with a circular saw, and glue it in place.

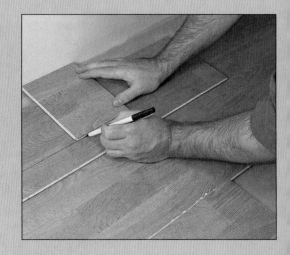

3 **Fit end boards** Since you can't use strap clamps or a tapping block to tighten up the joints on final boards and end planks (like the one shown), you'll need a special tool known as a pull bar. A pull bar is basically a piece of bar stock with the ends bent over in opposite directions. One end is hooked over the plank to be tightened, and the other is struck with a hammer to pull the plank against the adjoining piece, tightening the joint.

Working Around Obstacles

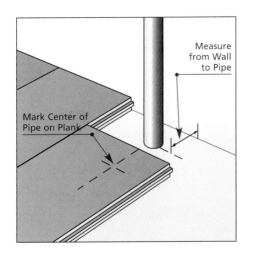

Measure from Wall to Pipe

Mark Center of Pipe on Plank

1 **Measure and mark** To allow laminate flooring to wrap around an obstacle (such as the pipe shown here), start by cutting a plank to length to fit between the last plank installed and the wall (subtracting ¼" for the expansion gap). Then measure and mark the center of the pipe on the end of the plank, *as shown.* Next, measure the distance from the wall to the center of the pipe, and transfer this measurement to the plank as well.

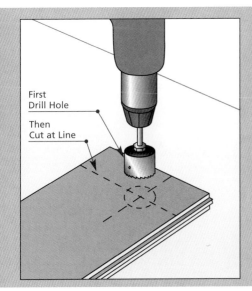

First Drill Hole

Then Cut at Line

2 **Cut and/or drill** Clamp the plank face-up on a sawhorse or other work surface, and drill a hole at the centerpoint you marked on the plank in Step 1. To allow for proper expansion, the bit needs to be ½" larger in diameter than the diameter of the pipe. Although a spade bit will work for this, you'll get a much smoother hole with less chip-out using a Forstner bit. A hole saw will also work well. After you've drilled the hole, draw a line through the center of the hole across the plank's width. Then cut the end of the plank off with a handsaw or power saw.

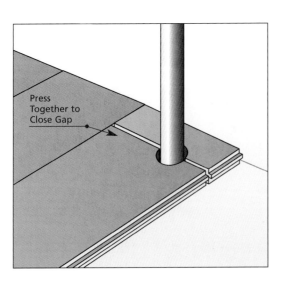

Press Together to Close Gap

3 **Fit and install** All that's left is to install the long piece as you would any other plank and then add the small cut-off end. Apply glue to the groove and the cut edge of the smaller piece and press it in place, making sure to insert spacers at the wall to ensure the correct expansion gap. Use a pull bar and a hammer to persuade the small piece to snug up to the larger piece, taking care not to disturb the pipe. Tip: To keep pressure on the cut joint, you may need to tap a carpenter's shim between the wall spacers and the small piece.

Finishing Up

1 **Clean up excess glue** As you lay the planks and tighten them up with clamps, you should experience quite a bit of glue squeeze-out. It's best to remove this immediately with a damp sponge (*as shown*) or a clean soft cloth. Have a bucket of clean water handy to rinse out the sponge or cloth periodically. After the entire floor is down and you've waited 12 hours for the glue to set up, go over the floor with a damp mop, using an ammonia-and-water mixture (½ cup of ammonia to 1 gallon of water) to remove any glue haze. If you find dried glue, go on to Step 2.

2 **Scrape glue** If you come across a large blob of semi-dried glue, use a plastic putty knife to scrape away the bulk of the blob. Don't use a metal putty knife, as it will damage the flooring. Scrubbing the remaining glue with a damp sponge or cloth will usually remove any residue. Another option is to use acetone to remove dried glue (check the manufacturer's directions before using this). Apply a small amount to a soft, clean rag, and scrub away the glue.

3 **Install thresholds** Most manufacturers of laminate flooring sell transition strips and molding to match their product. The T-molding shown here is just one of many styles available. It joins two areas of laminate flooring or other similar-height flooring. This type of transition is a two-part snap-in system that's easy to install. Following the manufacturer's instructions, cut both the metal track and threshold to fit the door opening. Screw the track to the floor, and press the threshold down into the track to snap it in place.

Shoe Molding

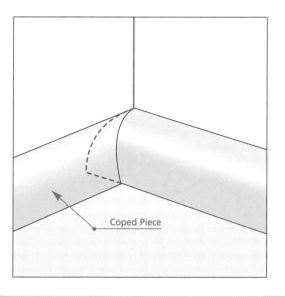

Coped Piece

Coped joint Once you've cleaned up the flooring and installed any thresholds, work around the room's perimeter, installing shoe molding to conceal the expansion gap. Although you can simply miter the joints where the molding meets at inside corners, a better method is to "cope" the joint. Coping a joint creates a very tight joint that won't show a gap—it's especially useful for out-of-square corners (like those in most homes). One piece of molding is cut square on the end and butted into the corner; the other is coped (*see below*) to wrap around it (*as shown*).

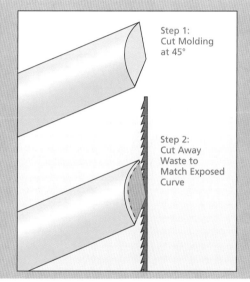

Step 1:
Cut Molding
at 45°

Step 2:
Cut Away
Waste to
Match Exposed
Curve

Coping a joint Here's how to cope a joint. Measure the length of the piece you'll need, then add the thickness of the molding to this measurement and cut a piece to length. (For example, if the wall is 80" long and you're using ¾" shoe molding, cut the piece 80¾" long.) Next, cut the end of the molding to be coped at a 45° angle with a handsaw and a miter box, or a chop saw. This will expose the molding profile. Now all you have to do is cut into the mitered end with a coping saw following the curved profile, *as shown.*

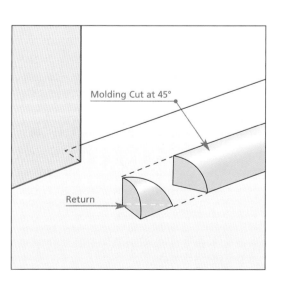

Molding Cut at 45°

Return

Returns A nice finishing touch that's commonly used by trim carpenters is to install a "return" on the end of a molding, *as shown*. Instead of leaving the end grain exposed, the molding is cut at 45°, and a small mitered piece is glued to the end. The molding effectively wraps around the end and "returns" to the wall. Cut the end of the molding at 45° and then make an opposite 45° cut on a piece of scrap molding. Then cut a piece off the scrap at 90° to create the return; glue this small piece to the mitered end of the molding.

Chapter 8
Resilient Flooring

Whenever I think of vinyl sheet flooring, an image pops into my head of the old "linoleum" we had in our kitchen when I was growing up. It was quite hideous—yellow with black spots—and virtually impervious to anything. (I'm sure that pattern is "in" again as a retro look.) And growing up with six siblings, believe me we hit it with everything: wagons, tricycles, jumping off the countertops, mixing mudpies, dropping pots, pans, and cutlery—you name it.

The resilient sheet flooring that's available today is available in an amazing array of colors and patterns (I'm sure I could even find that old yellow and black pattern). And just like the sheet flooring of days gone by, it's still impervious. What makes it even better now is that the standard no-wax finish it comes with is incredibly easy to clean and maintain.

Resilient flooring is the most common flooring in kitchens and bathrooms. It's relatively inexpensive and easy to install. An important installation note here: Unlike most flooring, which gets installed a piece at a time (like ceramic tile, strip hardwood flooring, and laminate flooring), sheet vinyl usually goes down in one piece. This means there's no margin for error. The only way to prevent mistakes with a single sheet is to make a template of the floor and use it to cut the flooring—don't even think about installing this type of flooring without a template.

In this chapter, I'll start by fully explaining the two common methods of installing resilient sheet flooring: full-adhesive and perimeter (*see the opposite page*). Then I'll show you how to make a template of the room so that you can cut your flooring to fit with confidence (*pages 84–85*). Then on to the first method of installation: perimeter. I'll go over everything from cutting the flooring and seaming it, to attaching it to the subfloor (*pages 86–89*).

After that I'll take you a step at a time through a full-adhesive installation, starting with cutting the tile and ending with installing transition strips (*pages 90–91*). Finally, there's a section on how to install cove base molding—the perfect trim for vinyl flooring (*pages 92–93*).

Types of Installation

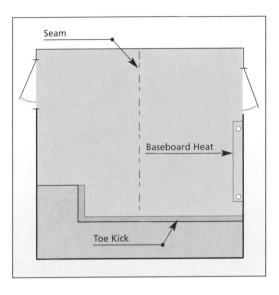

Full-adhesive When resilient sheet flooring is attached to the subfloor with flooring adhesive, it's referred to as a full-adhesive installation (*see the drawing*). A full-adhesive installation has a couple of advantages. First, since the entire floor covering is firmly glued to the subfloor, this type of installation is very durable. This is especially true if the sheet flooring is a single piece; it has no seams, so water, dirt, and dust can't sneak under it to weaken the glue bond. Second, being firmly attached to the subfloor helps prevent tears and rips that would otherwise occur if it weren't solidly held in place.

The downside to a full-adhesive install is that it's a real hassle to remove this type of flooring once the adhesive has hardened: A lot of back-breaking scraping is involved. Fortunately, if the flooring is firmly attached to the subfloor, you can usually lay a new covering directly over it.

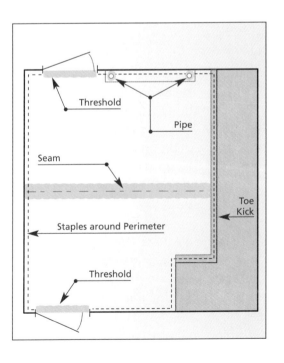

Perimeter with staples When sheet vinyl flooring is attached to the subfloor only around the perimeter (and at the seams, if applicable), it's referred to as a perimeter installation (*see the drawing*). Except for the areas around a seam, obstacles, and the thresholds, the only thing holding the flooring in place are staples around its perimeter. The rest of the flooring rests or "floats" on the subfloor. The advantage to this is that a perimeter install is much easier and not as messy as a full-adhesive installation.

A perimeter install works best with nonbacked solid vinyl flooring; it's flexible and can be stretched as it's stapled down. This makes it much more forgiving than paper-backed vinyl. Now for the bad news. Since most of the flooring isn't attached to the subfloor, it tends to tear or rip when abused—even high-heeled shoes can damage it. Also, since most perimeter sheet vinyl is stretched slightly as it's attached, a knife dropped in the kitchen can puncture the flooring, often creating a large rip.

Making a Template

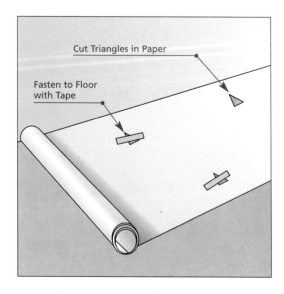

1 **Roll out paper; cut triangles** To make a template for installing resilient sheet vinyl, start by butting the edge of a roll of heavy paper (builder's paper or red rosin paper works great for this) into one corner of the room. Fasten the paper to the floor by first cutting small triangles with a utility knife in the paper as shown near the edges and throughout the paper at regular intervals. Then remove the paper triangles and press a strip of masking tape over each hole.

2 **Continue around perimeter** Next, continue rolling paper out along the perimeter of the room. Overlap the pieces 2", and fasten them together with strips of masking tape at the seams. Cut triangular holes in the paper as you did for the first strip, and fasten each piece to the floor as you work. Butt the edges of the paper as close as possible to the wall. If there's more than a ¼" space between the paper and the wall, trim the paper with scissors or a utility knife so it fits snug.

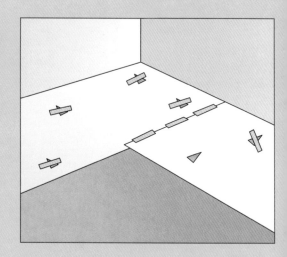

3 **Use a square to mark obstacles** Whenever you encounter obstacles along the perimeter of the room (such as the pipe shown in the drawing), cut the paper and fasten it temporarily to the floor on both sides of the obstacle. Then use a framing square to measure the distance from the wall to the obstacle. For round obstacles like pipe, measure to the center. You'll use this measurement in the next step to make a template from a separate piece of paper.

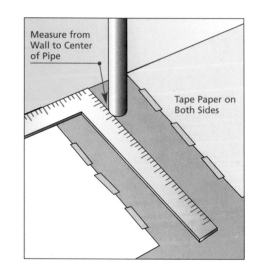

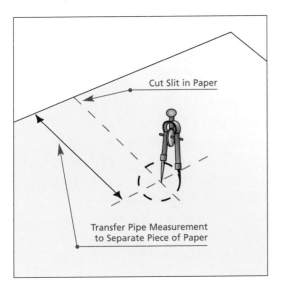

Cut Slit in Paper

Transfer Pipe Measurement to Separate Piece of Paper

4 **Transfer measurements to template** Next, cut a piece of paper to span the gap between the paper fastened to the floor so it'll overlap the pieces 2" in each direction. Use the framing square to draw a line perpendicular to the edge of the paper, and measure and mark the location of the obstacle (in this case, the center of the pipe). Then use a utility knife or scissors to slit the paper along the perpendicular line, and cut out the obstacle area on paper.

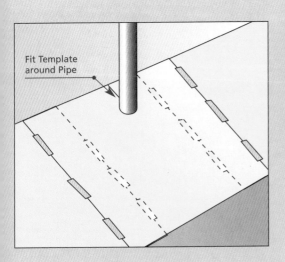

Fit Template around Pipe

5 **Test the fit** Now slide the template around the obstacle and test the fit. If it's too snug, trim the hole and test again. If it's too loose or it's in the wrong position, make another template—it'll take only a few minutes, and it's worth the time. Once you're satisfied with the fit, tape one edge of the template to the paper on the floor. Then lift the opposite edge and remove all the tape underneath that you used to temporarily hold the pieces to the floor. Press the template flat and fasten it to the remaining floor piece

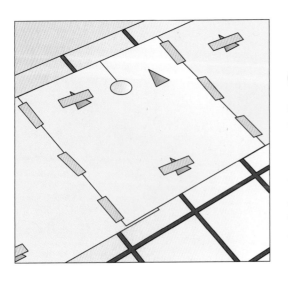

6 **Roll up and transfer to flooring** When you're done making the template, remove all of the tape covering the triangle cutouts and carefully lift it off the floor and roll it up. Transport it to the room where you've laid out the sheet flooring, and set it on the flooring. Position the template carefully so that you'll end up with the pattern you want. Keep the template an inch or two away from the factory edge so that you can trim it to fit the room. Use the same triangle cutouts to fasten the template to the flooring.

Perimeter Installation

Resilient sheet vinyl that's installed with the perimeter method is fastened to the subfloor with staples around the perimeter. The rest of the flooring (with the exception of a few special areas) is not attached to the floor—it simply rests on top of it. Any dirt or particles that are trapped under this are free to move around over time. Combine this with constant foot traffic, and you've got the potential for some grinding to take place. With this in mind, it shouldn't come as a surprise to you when I say that the preparation of the subfloor is critical to a successful installation. Not only must the floor be smooth and flat, it also has to be spotlessly clean.

If you'll be joining pieces together, you'll discover that the installation sequence may seem a bit odd. You might think you'd staple the pieces around the perimeter and then go back and join the pieces at the seam. But if you did it this way, you may end up with a gap at the seam. It's much safer to install it the other way: seams first, and then the perimeter.

1 **Roll up template** Once you've made a paper template for the room (*see pages 84–85*), carefully roll it up and transport it to the room where you'll be cutting the flooring. Take your time here to make sure the template doesn't get torn or creased as you move it. Although you can leave the masking tape in place that you used to fasten it to the floor with, it's best to remove these, as they tend to stick to the template when you roll it up.

2 **Unroll flooring** Professional flooring installers often take the template they've made of your floor back to their shop. There they've got a huge space where they can unroll the flooring and then place the template on it for cutting. Unfortunately the rest of us have to make do with what's available. Since the living room or family room in many homes is the largest room, consider unrolling your flooring there. It may be necessary to move some furniture, but it's well worth the effort. If you try to work on the flooring and it's not laid flat, chances are high that you'll end up with cut flooring that doesn't fit.

3 **Overlap pieces and tape** For floors that require pieces to be joined, overlap the pieces and tape them together. Position the pieces so that the pattern flows perfectly from one piece to the other. Be careful here to observe small details like shading. The flooring shown imitates ceramic tile. To give the illusion of 3-D tiles, two edges of each dark green tile are dark and the opposing two are lighter. It would be easy to overlook this until the flooring was in place and attached to the floor.

4 **Position template** The next step is to unroll the template onto the flooring and temporarily fasten it in place with masking tape. This is another place to take your time. Make sure that it's oriented on the flooring properly. If the new flooring doesn't have a pattern, there's nothing to worry about. But if it does, it's really easy to get confused here. Double- and triple-check that it's correct before proceeding to the next step.

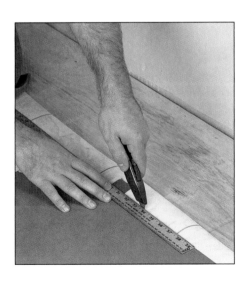

5 **Cut perimeter** Once you're sure that the template is fastened to the flooring correctly, you can cut it to match the template. Start by making the straight perimeter cuts with a utility knife, using a metal straightedge as a guide. Slide a scrap of plywood under the area you're cutting, to protect the existing floor. Use firm, steady pressure as you cut. Most tile will cut cleanly in a single pass, as long as the blade is sharp. You can also cut a light taper using the straightedge by angling it slightly to match the taper.

6 **Make obstacle cuts** After you've cut the perimeter, use a utility knife to cut holes and curves to fit around obstacles. Don't forget to cut the slits necessary for the flooring to be slipped around the obstacle. For intricate curves, try making a series of light cuts instead of one heavy one. You'll find it's easier to make gradual curves this way. If you're cutting around a square or rectangular obstacle, use a straightedge to guide the cut.

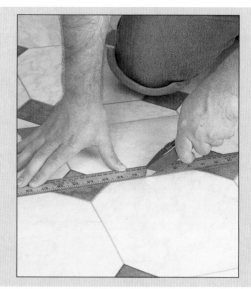

7 **Cut seam if necessary** At this point in time, you can remove the template. I'd suggest saving it in case you want to install different flooring in the future—it doesn't take up much space, and it'll save you a lot of work next time. If the flooring you're installing is more than one piece, it's time to cut the seams. With a scrap of plywood under the seam to protect the flooring, use a sharp utility knife guided by a metal straightedge to cut through both layers of flooring at the same time. If your flooring has imitation grout lines, try to split the grout line in half as you cut.

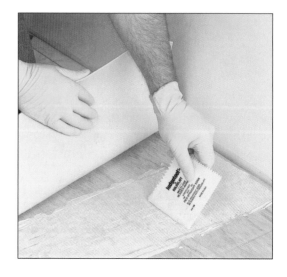

8 **Apply adhesive under seam area** Move the flooring into the room where it will be installed. If your flooring is in more than one piece, you'll need to apply adhesive under the seam area. With the pieces in place on the floor, fold back one piece at a time near the seam and apply adhesive to the subfloor with the appropriate-sized trowel. Make sure to use the adhesive recommended by the manufacturer. Repeat for the other half of the seam.

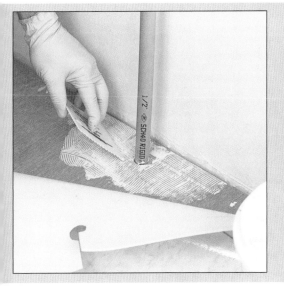

9 **Press seam with roller** After the adhesive is applied, fold back each section of flooring and press the edges together to form a tight seam. Use a small hand roller, rolling pin, or laminate roller to press the flooring firmly onto the adhesive. Wipe up any squeeze-out with a soft cloth dampened in solvent. Strips of duct tape applied across the seam every 4" to 6" will help hold it together until it has set up. Following the manufacturer's instructions, allow the adhesive to dry and then apply the appropriate seam sealer (*inset*).

10 **Apply adhesive around obstructions** Even if you don't have seams, you'll still need to apply adhesive to a couple of areas with the appropriate-sized trowel. First, you should apply adhesive around any obstruction. This will keep the flooring flat and secure without having staples visible. Second, it's a good idea to apply adhesive to the first 6" to 8" around a door threshold. Although this area is stapled as well, the adhesive helps prevent the flooring from stretching unduly under the constant barrage of foot traffic.

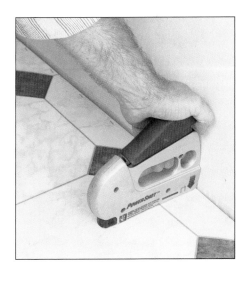

11 **Staple around perimeter** Now comes the fun part. Starting in one corner, pull the flooring tight against the wall and staple it to the subfloor. I like to use a lot of staples—one about every 2". Continue working around the room, pulling and stapling. You'll be surprised how much nonbacked vinyl will stretch. Keep the staples as near to the wall as possible. They'll be covered later when you install cove base molding (*see page 92*), other trim, or thresholds. If a staple doesn't go all the way in, give it a whack with a hammer.

Full-Adhesive Installation

Although I have perimeter-installed sheet vinyl in my kitchen, it still makes me nervous. Call me old-fashioned, but I'd feel better if it were stuck solidly to the subfloor with adhesive. I realize that vinyl flooring is designed to stretch, but when I stop suddenly walking to the sink, I can feel the flooring move beneath. Now I realize I'm a big guy, but movement like that just doesn't seem right. The next floor that goes down will be glued.

Yeah, sure it's messy, but what home-improvement job isn't? Installing full-adhesive sheet vinyl is very similar to installing perimeter vinyl flooring. The big difference is at the end, when you go to fasten it to the subfloor—instead of staples, the entire subfloor gets covered with adhesive. The downside to this, as I mentioned earlier, is that removing this type of flooring (in the future) is hard work.

1 **Cut the flooring** Using the method described on pages 84–85, make a paper template of the room in which you'll be installing the flooring. Roll up the template and move it into the room where you've unrolled the flooring. Fasten it to the flooring by covering the triangular cutouts with masking tape. Cut around the perimeter with a utility knife, guided by a metal straightedge. Then make any cuts necessary to wrap the flooring around obstacles. Remove the template and move the flooring into the room where it'll be installed.

2 **Pull back and apply adhesive** Position the flooring on the subfloor, taking care to slip it around obstacles. If you're working with one piece, pull one side back toward the center and apply flooring adhesive with the recommended notched trowel—usually 1/8". Then carefully fold it back into position. Repeat this process for the other half. If you're working with multiple pieces, see Step 4.

3 **Roll it with a floor roller** Before you begin pressing the flooring into the adhesive with a flooring roller (rented), go around the room to make sure the flooring is positioned properly. Once you start pressing it in place, it's a lot more difficult to lift and reposition. Begin rolling in the center of the room, working your way toward the wall. This pushes out air bubbles so they can escape and also moves any excess adhesive to the edges, where it can be removed.

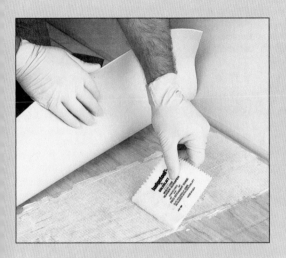

4 **Do the other section** If you're working with multiple pieces, use the same process of folding the flooring over onto itself, except do this for each piece. Have on hand a soft cloth dampened with solvent to clean up the inevitable mess. It's best to roll and press one piece in place at a time. This gives the other piece a solid edge to butt up against—there's just a lot less slipping and sliding this way.

5 **Transition strips** To protect the edges of the sheet flooring at the door thresholds, install metal transition strips. To do this, measure the door opening and cut the transition strip to length with a hacksaw. Then attach the strip to the subfloor with the nails provided. Home centers carry transition strips in standard door widths in a variety of colors to handle the transition from just about any type of flooring to another.

Cove Base Molding

Vinyl cove base molding is the perfect trim for a resilient floor. It's easy to work with, goes on fast, and comes in a wide variety of colors to complement almost any flooring pattern. Cove base molding is available in rolls and strips. Roll molding can save you a lot of time making seams and is the way to go with large rooms. Although strip molding requires seams, its short lengths make it a lot easier to work with. When you buy molding, pick up a couple of extra strips for future work and repairs; you'll be glad you did.

All it takes to work with cove base molding is a sharp utility knife, a notched trowel, and some cove base adhesive. Whenever possible, try to use water-based adhesive—it's much easier to clean up than the solvent-based varieties. Although most cove base molding is flexible enough to wrap around corners, it looks a lot better if you cut and miter the joints (*see Steps 4 and 5 on the opposite page for more on this*).

1 Apply adhesive Applying adhesive to cove base molding is the perfect time for the "back-buttering" technique, where you apply the adhesive to the back of the molding instead of onto the wall. The reasons for this are simple: It's both awkward to apply it to the wall and difficult to keep it at the correct height. Apply the adhesive to the back of the molding with a notched knife or trowel. To avoid squeeze-out, keep the adhesive about 1/2" from the top edge.

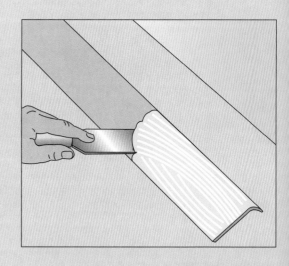

2 Set in place Read and follow the manufacturer's instructions on how much time (if any) you should let the adhesive set up (or get tacky) before installing the molding. When it's ready, carefully position the molding near the wall with the curved lip at the bottom resting on the floor. Then tilt the molding up until it rests against the wall. Press it firmly in place with your hand, working from one edge of the strip to the opposite edge.

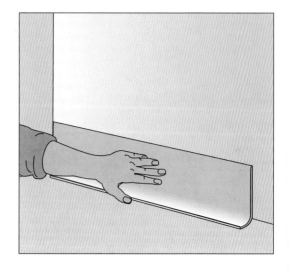

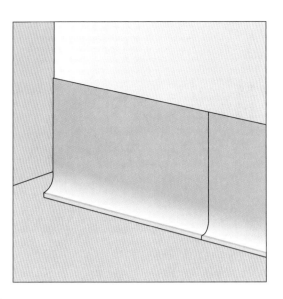

3 **Butt joint** The most common way to join strips of cove base molding together is with a butt joint. To do this, simply butt the edges of the strips together and hold the seam tight for a few seconds. A strip of duct tape across the seam will help it stay flat until the adhesive sets up. Another option, if you've got a steady hand, is to cut opposing bevels on the edges of the strips to be joined. This is similar to a scarf joint used with wood trim. If done properly, the seam will be virtually invisible.

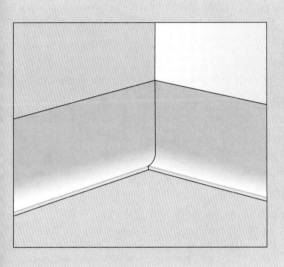

4 **Inside corner** As I mentioned earlier, cove base molding is flexible enough to wrap around most corners—either inside or outside corners. The trouble with this is that there will always be a gap at the top where it doesn't conform perfectly to the edge of the corner. These gaps never fail to collect dirt and dust. To get around this, consider mitering the pieces of base molding just as you would with wood trim. A sharp utility knife, a steady hand, and some practice is all it takes.

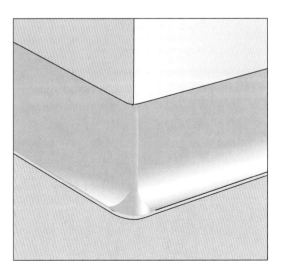

5 **Outside corner** Just as you can miter the ends of cove base molding to meet in a corner, you can miter them to fit together at an outside corner. This does take some practice, and you might find drawing a line on the cove base to serve as a reference will help you make a straight cut. For the tightest possible joint, apply a thin coat of adhesive on the mitered ends before pressing the pieces together. Wrap a piece of duct tape around the corner to hold it place until the adhesive sets up.

Chapter 9
Wood and Vinyl Tiles

Ease of installation. That's the first thing that pops into my mind when I think about vinyl or parquet tiles. That's because of all the flooring options, they are one of the easiest to install. Unlike sheet vinyl flooring and carpeting, where you're faced with struggling with large, heavy, and awkward sheets, 12" squares seem blissfully workable. Small square tiles are a lot easier to maneuver, especially when it comes time to work around an obstacle like a set of pipes or a piece of conduit.

Unfortunately, both wood and vinyl tiles have suffered from a poor reputation for the last decade or so. One of the main causes of this is self-adhesive tile. It seemed like such a good idea—just peel off the backing and press the tile in place for a new floor—that many homeowners gave it try, only to be sadly disappointed. The problems were many. To allow the tiles to conform to uneven floors, manufacturers made the tiles thin. This made them prone to cracking, chipping, and denting. Also, adhesives at the

time weren't as strong as today's glues, and the bond between floor and tile often failed.

But when installed properly, both parquet and vinyl tiles can last a long time. Want a reference or two? Check out the floor in your local school, library, hospital, or other institution. Odds are, it's vinyl tile. Vinyl tile is still the choice for many buildings that are exposed to constant traffic. Doesn't your kitchen or bathroom qualify as one of these?

In this chapter, I'll show you the correct way to install both parquet and vinyl tile. I'll start with guidelines (*opposite page*) and jump right in with how to lay down vinyl tile so it'll last a long time—including how to handle border tiles and irregular tiles (*pages 96–99*). Although self-adhesive tiles still don't stand up as well as full-adhesive tiles, I've included a short section on how to install them (*page 100*). Finally, I'll take you step-by-step through how to install wood parquet tiles (*pages 101–103*).

Guidelines

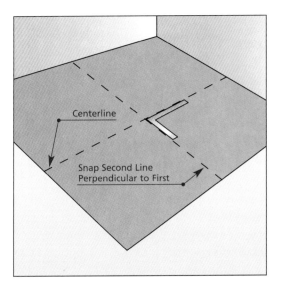

Centerline

Snap Second Line
Perpendicular to First

1 **Establish layout** An accurate set of reference lines will make any vinyl or parquet tile installation go a lot smoother. To do this, start by measuring and snapping a chalk line down the center of the room. Then use a tape measure to locate the center of the line you just snapped. Next, use a framing square to lay out a line perpendicular to the first line. Align your chalk string with the line you just drew, and snap a line perpendicular to the first line.

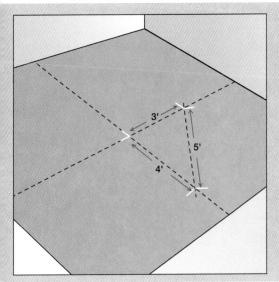

3'

5'

4'

2 **Check guidelines** To check to make sure the lines you just snapped are truly perpendicular, use a 3-4-5 triangle: Measure and mark a point 3 feet from the inter-section of the lines. Then measure and make a mark 4 feet from the intersection on the adjacent side (*see the draw-ing*). Now measure from the 3-foot mark to the 4-foot mark. If the lines are truly perpendicular, it will be exactly 5 feet. If it's not, adjust one of the lines until this distance is exactly 5 feet.

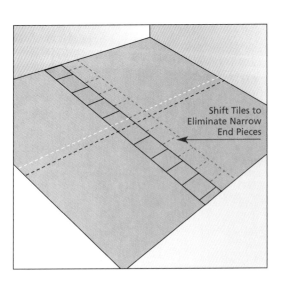

Shift Tiles to
Eliminate Narrow
End Pieces

3 **Adjust if necessary** To prevent narrow tiles at the perimeter of the room, temporarily set out a row of tiles, starting at the centerpoint and working toward the walls, *as shown*. If you find a narrow gap between the last full tile and the wall on either end, shift the appropriate centerline to eliminate it. Repeat this process for the opposite direction to make sure you don't have any nar-row tiles on the remaining walls.

Installing Vinyl Tiles

As I mentioned earlier, one of the things I like best about vinyl tiles is how easy they are to install. Besides a rented flooring roller that you'll use to firmly press the tiles into the flooring adhesive, you won't need any specialized tools. Also, vinyl tiles are one of the easiest flooring materials to cut—all it takes is a utility knife. Working around obstacles is a snap since you're working with small pieces. And cleanup is a breeze if you choose a flooring adhesive that's water-based.

   You'll find that it's more work to prepare the floor than it is to install the tile. That's one catch to vinyl tiles. Since there are so many seams, the subfloor really needs to be flat and free of debris. Even a small piece of sawdust trapped under a vinyl tile can cause the tile to crack over time or weaken the adhesive bond. Take care to scrape the subfloor to remove any old residue and carefully vacuum the floor before applying adhesive.

1 **Spread adhesive** Starting at the intersection of the reference lines, apply flooring adhesive to the subfloor with a V-notched trowel. Follow the manufacturer's suggestions as to notch size; you'll find this information on the adhesive can. Hold the trowel at approximately a 45° angle as you spread the adhesive, and apply it until you reach one wall. Take care not to obscure with adhesive the reference lines that you'll use to lay tile.

2 **Fill a quadrant** Carefully set a row of tiles along one of the reference lines. Don't slide the tiles into place; instead drop them into position. Then set a row of tiles perpendicular to the first row you laid down. Next, fill in the quadrant by filling in between the two outer rows. Note: Most tiles have an arrow printed on the back to indicate which direction they should be laid. Make sure that all the arrows are facing the same direction as you install them.

3 **Roll tiles** To create the best bond between the subfloor and the tile, the tiles need to be firmly and evenly pressed into the flooring adhesive. Without a doubt, the best tool for this job is a flooring roller, rented from a nearby rental center. Thin tile can be pressed with a 75-pound roller; thicker tiles (such as rubber tile) are best pressed in place with a 100-pound roller. If you're doing only a small area, you can get by in a pinch with a rolling pin; keep your weight over the pin as you roll, for maximum pressure.

4 **Clean off adhesive** As you press the tile into the adhesive with the roller, you're likely to experience quite a bit of squeeze-out. The flooring adhesive will ooze out between the tiles and create a mess. One way to minimize this is to clean the roller frequently to prevent it from spreading the adhesive around. Whenever you stop to clean the roller, use a cloth or sponge dampened with solvent to remove any adhesive that's squeezed out between the tiles. (Wear disposable gloves to keep your hands free of this sticky stuff.)

5 **Repeat** Once you've got one quadrant laid down, pressed, and cleaned up, you can move on to the next. As you fill in the three remaining quadrants, take care not to step on the freshly laid tile until the adhesive has had a chance to set up (check the label on the adhesive for instructions and set up times). Once all the full tiles are down, go around the perimeter and measure, cut, and install the border tiles (*see page 98*).

Border Tiles

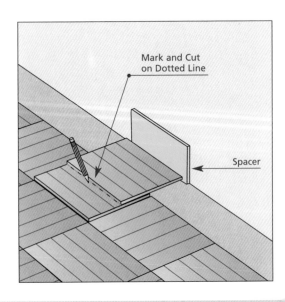

Mark and Cut
on Dotted Line

Spacer

1 **Measure** After you've installed all of the full tiles possible, you can begin work on the border tiles. To mark a tile for cutting, start by placing a tile on the full tile nearest the wall. Then place a ⅛" spacer against the wall, place a "marker" tile on top of the tile to be cut, and slide the marker tile over so it butts up against the spacer. Next, using the edge of the marker tile as a guide, draw a line on the tile. Now you can cut the bottom tile to fit. Install it as you would a full tile, except take care to leave the gap between the wall and the tile.

2 **Corner tiles** Border tiles that wrap around a corner can be marked and cut in much the same way as the simple border tile shown in Step 1. The only difference is that you need to mark the bottom tile twice—once for each side of the wall you're going around (*see the drawing*). Here again, it's important to insert a ⅛" spacer between the tile and the wall when marking the tile to leave an expansion gap.

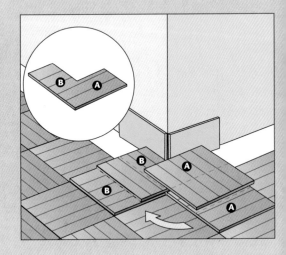

3 **Border trim** Although you can use the same techniques shown in Steps 1 and 2 for parquet border tiles, another option that adds a nice decorative touch to the floor is to use border trim. This can be a narrow strip of matching or contrasting solid wood that's cut to fit around the perimeter of the room. Inside and outside corners are mitered for a smooth transition (*see the drawing*). Just like the tiles themselves, the border trim can be glued to the subfloor with flooring adhesive.

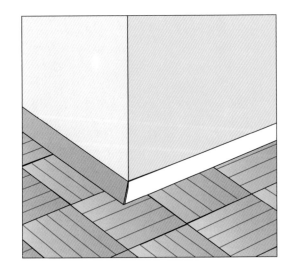

Irregular Tiles

Paper template Probably the most foolproof way to mark an irregular tile for cutting is to make a paper template like the one shown here. The idea is that you make mistakes on paper and not on tile. Start by cutting a piece of paper the same size as the tile—builder's felt or red rosin paper (*as shown here*) works great for this. Then carefully measure and mark the obstacle location on the paper. Next, cut the template to size and check the fit around the obstacle. Make any necessary adjustments before transferring the template onto the tile for cutting.

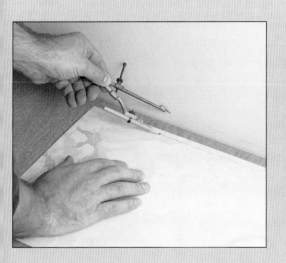

Compass Another method that I often use to mark tiles to be cut is to use a compass to "scribe" or copy the irregularities of a wall onto a tile. To do this, place the tile as close as possible to the uneven wall. Then open a compass up far enough so that the points span the largest gap between the tile and the wall. Now with the point of the compass against the wall and the pencil on the tile, guide the compass along the wall. The pencil will copy the irregularities directly onto the tile.

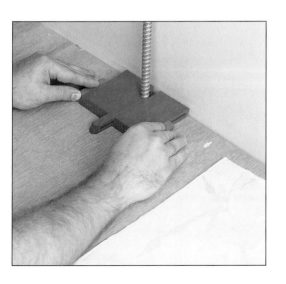

Contour gauge If you've got a contour gauge handy, it's one of the best ways to copy a complicated profile (such as a door casing) onto a tile to be cut. Press the fingers of the gauge firmly against the profile to be copied. The fingers will conform to the profile to produce both a negative and a positive. Either can be used to trace the profile onto the tile. Just make sure to measure the location of the profile carefully so you can transfer the profile to the tile at the appropriate spot.

Self-Adhesive Tiles

What could be simpler than installing self-adhesive or "peel-and-stick" tiles? You peel off the protective paper backing and press the tile in place. It really is that easy. Of course, you'll still need to prepare the subfloor. It must be flat, level, and free from dirt. Cleanliness is extremely important with self-adhesive tiles because even the tiniest bit of dirt can contaminate the bond, resulting in a weakly attached tile. Also, just like vinyl tiles, most self-adhesive tiles have directional arrows on their backs to let you know how they should be installed.

There are a couple downsides to self-adhesive tile, however. First, most of these tiles are very thin so that they can easily conform to the floor to get the best possible bond. Because of this, they tend to crack and dent easily. Also, they have a well-deserved reputation for losing their bond over time. The water and detergent from repeated scrubbing sneak in through the seams and degrade the bond, eventually causing it to fail.

1 Peel backing off To install a self-adhesive tile, start by peeling off the protective paper backing. If you need to cut a tile, you'll find that most self-adhesive tiles are thin enough to be cut with a pair of heavy-duty scissors. Don't remove the paper backing and then cut the tile; instead, leave it in place and make your cut. This way, the tile won't stick to the scissors. Since you'll be generating a lot of waste with the backing, it's a good idea to have a helper handy to collect and dispose of the backing.

2 Press and stick To lay self-adhesive tiles, position them carefully along your reference lines. Set them in place, and press down firmly with the palms of both hands. Place adjacent tiles so they butt firmly against one another. Unlike vinyl tiles glued to a floor, you can't easily lift up a self-adhesive tile if it goes down crooked. If you do need to reposition one, slide a wide-blade putty knife under the tile to help break the adhesive bond. Discard this tile and start with a fresh one. As with full-adhesive tile, you'll get a better bond by pressing the tiles with a flooring roller after setting them all in place.

Parquet Tiles

Parquet tiles are sort of a hybrid of vinyl tiles and hardwood flooring. They offer both the convenience and ease of installation of tiles and the warmth and beauty of wood. You can purchase parquet tiles either unfinished or prefinished. Each has its advantages. Unfinished tiles can be stained to match surrounding woodwork, but they require additional work. A prefinished tile floor, on the other hand, is ready to use as soon as it's down. The downside to this is that you have to live with the colors that the manufacturers offer.

Tip: Consult a variety of flooring suppliers when looking for prefinished tiles. Supply houses sell only certain brands, with set color palettes. You may be able to find what you're looking for at another supply house that carries a different brand. Most parquet tiles come in 12" squares, in thicknesses varying from ⁵⁄₁₆" up to ¾". By shopping around, you may be able to find different sizes and shapes—octagons and rectangles are common.

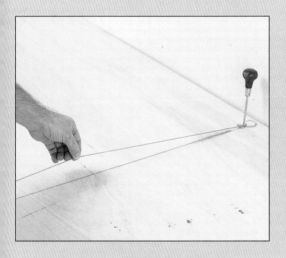

1 **Snap chalk lines** As with any tile installation, the first step for laying parquet tiles is to establish a set of perpendicular layout lines. (*See page 95 for step-by-step directions on how to do this.*) Since parquet tiles are more difficult to cut than vinyl, and can virtually self-destruct if cut into narrow strips, take the time to make a trial run (*see Step 2*) to eliminate the possibility of having narrow border tiles.

2 **Trial run** Make a trial run with the parquet tiles in one of the quadrants both to check for narrow border tiles and to get used to their interlocking system. Most parquet tiles have two adjacent edges with tongues and two adjacent edges with grooves. As you mate the tiles together so that tongues on one tile fit into the grooves on another, you'll create a basket-weave pattern.

③ Spread adhesive Beginning at the intersection of the reference lines, spread flooring adhesive in one of the quadrants using a V-notched trowel. Check the adhesive label for notch size and application instructions. Make sure that as you spread the adhesive, you don't obscure the reference lines. Spread only as much adhesive as you can work before it sets up; check the adhesive label to see whether you need to allow it to get tacky before installing tiles. If you're using self-adhesive parquet tiles, you won't need to apply any adhesive.

④ Set first tiles Lay the first tile in the corner of the intersection and align the edges of the tile, not the tongues, with the reference lines. Insert the tongues of the next tile into the grooves in the first tile, taking care to make sure it interlocks fully. As with ceramic tiles, you may find it helpful to install battens (*see page 50*) to help prevent the parquet tiles from sliding around in the adhesive. Avoid walking on the newly installed tiles, but if you must, lay down a sheet of plywood first to distribute your weight evenly.

⑤ Bed the tiles After you've filled in one complete quadrant with full tiles, the next step is to "bed" the tiles—that is, press them firmly into the adhesive to create the best possible bond. Here again, the tool of choice is a flooring roller. I'd suggest the lighter, 75-pound version for parquet tile. It has less of a tendency to dent or damage the tile. When you're done, check the floor to make sure no adhesive has squeezed out between the joints. Use a cloth dampened with solvent to clean up any excess.

6 **Border tiles** The tiles that run along the perimeter of the room must be cut to fit. Two things here. First, many parquet tile manufacturers embed a corrugated metal strip in the back of each of the four quadrants of a tile to hold the strips together. Make sure you use a non-carbide-tipped blade to cut these tiles: If carbide-teeth hit metal, they'll fracture and send tiny bits of shrapnel flying. Second, you need to leave an expansion gap between the tile and the wall—usually ¼" to ⅜"—check the tile installation instructions for the proper gap for your tile.

7 **Fitting border tiles** Fitting parquet tiles to span the gap between a wall and the nearest full tile is similar to the technique shown on page 98 for vinyl tiles. The only differences are that you'll need to use a larger spacer, since the expansion gap is bigger, and when you position the tiles for marking, align the edges of the tiles, not the tongues. (*For tips on cutting parquet tile, see the sidebar below.*)

CUTTING PARQUET TILES

There are number of tools that you can use to cut parquet tile. Since many of the cuts you'll need to make won't be straight—they'll be tapered to match a wall, or curved to go around an obstacle—I suggest a saber saw. Clamp the tile to a sawhorse or other work surface so the edge to be cut extends out past the sawhorse. Then make the cut while supporting the waste piece (*left photo*). To prevent chip-out, apply a strip of masking tape over the line to be cut. Redraw the line on the masking tape and make the cut (*right photo*). The tape supports the fragile wood edge to minimize splintering.

Chapter 10
Flooring Repairs

Even the toughest of flooring materials will eventually surrender to the constant onslaught of abuse from daily traffic and heavy furniture. Muddy boots, kids, pets, and mobile furniture such as a microwave cart or a wheeled desk chair all take their toll. Rips, tears, cracks, and stains are common. If the damaged area is small, it can be patched as long as you can find suitable replacement materials. Finding these is often one of the largest challenges you'll face. If you can't find appropriate patch material or if the damaged area is large, you may be better off removing the old flooring and installing a new floor—or simply covering the old flooring with a new layer.

In this chapter I'll show you how to make repairs to the most common types of flooring. I'll start with a pet peeve of mine—squeaky floors. It never ceases to amaze me how many people accept squeaks in floors as if they couldn't be fixed. They can. Check out pages 105–107 to learn a wide variety of ways to eliminate them.

Next, I'll go over ways to repair carpet: everything from small repairs with a special tool called a tuft setter, to removing larger sections and inserting an almost invisible patch (*pages 108–109*). Then on to patching hardwood strip flooring. I'll cover two methods: a rectangular patch that's fairly easy to do, and the more challenging but less noticeable staggered patch (*pages 110–113*). Since hardwood flooring is especially long-wearing, I've included a section on how to refinish a floor (*pages 114–115*) and how to remove surface stains (*page 116*). If your wood parquet tile floor is damaged, I'll show you how to repair that as well—either by replacing single strips (fillets) or by removing and replacing entire tiles (*page 117*).

How-to repair instructions for resilient sheet flooring, including a nifty tip to make seams both strong and nearly invisible, are on pages 118–119. This section is followed by instructions on making repairs to vinyl tiles (*pages 120–121*). **Safety note:** Before attempting any repair on these types of flooring, read the introductory text on page 118 regarding asbestos.

The remainder of the chapter covers ceramic tile. Included are directions on how to remove stains (*page 122*), replace grout (*page 123*), and replace damaged tiles (*pages 124–125*).

Squeaky Floors

I don't know about you, but squeaky floors really annoy me. I've lived in old houses where you can tell where someone is in the house just by listening to the squeaks. Some may consider this charming; to me it's a floor begging for repair. Fortunately, most squeaks can be eliminated with a little detective work and some simple techniques.

The main reason a floor squeaks is because there is unwanted movement between the subfloor and the joists or between the subfloor and the flooring. Another common cause is a floorboard that has worked loose and is rubbing on a nail or screw when someone presses on the board by walking over it. Once you've located a squeak, the first thing to do is inspect the floor from beneath, if possible. Have a helper walk across the floor as you watch the floor from below. Mark any areas where you see movement, and use one of the methods shown on the following pages to eliminate the movement.

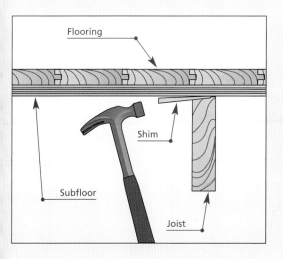

Wood shim Minor squeaks can often be eliminated by inserting a carpenter's shim between the subfloor and the floor joist. Once you've located the area with movement, gently tap in a shim with a hammer. Stop as soon as it becomes friction-tight—if you drive it in too far, you'll only increase the separation between the subfloor and joist and intensify the problem. Have a helper handy to test the floor. If it still squeaks, drive the shim in a bit further and retest. Persistent squeaking may require a more sophisticated repair (see below).

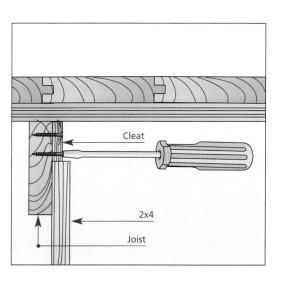

Cleats Another simple way to stop squeaks is to shore up the joist with a cleat. To do this, first cut a scrap wood cleat—a short length of 1×4 or 1×6 will do. Butt the cleat up against the floor, as shown, and press it firmly in place by wedging a long 2×4 between the bottom edge of the cleat and the floor. Tap the 2×4 as necessary to push up on the cleat until the gap and squeak are gone. To hold the cleat in this position, screw it to the joist. Now remove the 2×4 and test.

Steel joist bridges Supporting joists so they don't move or twist when weight is placed on them from above is a classic way to reduce squeaks. One of the simplest ways to do this is to install prefabricated steel bridging between the joists. Steel bridging can be found at most home centers to fit between joists spaced at 12", 16", and 24". To install a steel bridge, hammer the straight-pronged end into a joist near the top. Then pound the L-shaped end into the adjacent joist near the bottom. Alternate bridging to form a crisscross pattern.

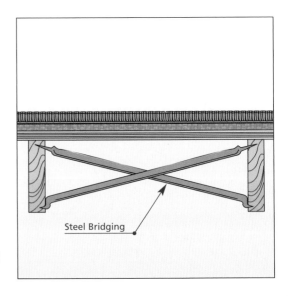

Steel Bridging

Wood bridges Occasionally, you'll discover a squeak that's between the joists. The solution here is to bridge the space between the joists with a cleat and then insert shims between the bridge and the subfloor as you did on page 105. Cut a 1×4 or 2×4 cleat to fit between the joists, and nail it in place. Then gently tap in shims to eliminate movement. Tip: Apply a bead of construction adhesive to the shim before installing it—this will further reduce movement and will hold the shim in place over time.

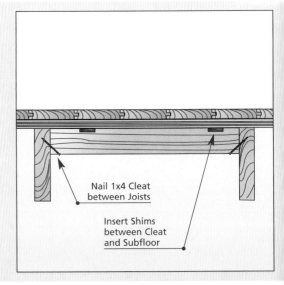

Nail 1x4 Cleat between Joists

Insert Shims between Cleat and Subfloor

ANTI-SQUEAK HARDWARE

For really persistent squeaks, consider one of the many types of anti-squeak hardware that you install from under the floor. (You can find these at most hardware stores and home centers.) The hardware shown here locks the offending piece of flooring to the joist so that it can't move anymore. This is done by screwing a stout bracket to both the joist and the subfloor. The entire installation takes less than a minute. Other systems use a metal bracket that slips over the joist and is pulled up toward the subfloor. This is done by tightening a nut that threads onto a rod screwed to the subfloor.

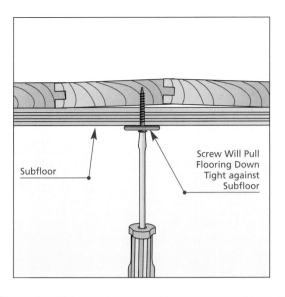

Subfloor

Screw Will Pull Flooring Down Tight against Subfloor

Screw from below For squeaks caused by flooring that has buckled over time, you can pull the flooring tight against the subfloor by screwing into it from below. Make sure the screws you choose will stop at least ¼" from the surface of the flooring. Since flooring of this type is usually hardwood, drill an appropriate-sized pilot hole. Use a large washer (such as a fender washer) to distribute the pressure around the screw, and apply a bit of wax to the screw threads to make it easy to install.

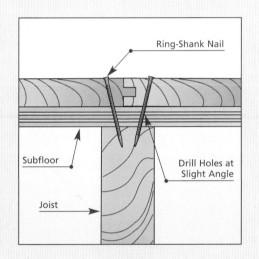

Ring-Shank Nail

Subfloor

Joist

Drill Holes at Slight Angle

Surface nailing Another option for dealing with strip hardwood flooring that squeaks is to force the strips flat against the subfloor by nailing them from above. Obviously, this is more noticeable than screwing the flooring in from below, but sometimes the floor isn't accessible from below (as when the ceiling below is finished). In this case, use a stud finder to find the nearest joist, and drill pilot holes for nails. Drill the holes at opposing angles (*as shown*) to lock the strips in place, and use ring-shank nails for the best possible grip.

TRIM-HEAD SCREWS

Squeaks in carpeted floors can often be eliminated in a manner similar to surface nailing (*shown above*). But instead of hammering nails and flattening the carpet, try using a special type of screw, called a trim-head screw. The head of a trim-head screw is much smaller than a standard screw head and can countersink itself, much like a finish nail. As with nails, drive in the screws at opposing angles to lock the floor to the joist. Since the screw countersinks itself, you'll never know it's there—except that the squeak is gone.

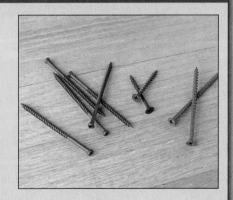

Repairing Carpet

Although many of today's carpets are stain-resistant, you'll occasionally encounter a material that the carpet just can't resist. If a professional carpet cleaner can't remove the stain, it's time to consider removing the offending area and replacing it with a patch.

The problem, of course, is finding a piece of carpet for the patch that will match the original carpet. If you're a pack rat like me, you probably have a remnant somewhere (the only problem may be finding it). If you don't have a leftover piece, contact several carpet suppliers to see whether they can help. When all else fails, you may be able to cut a small a section of the original carpet out of a closet or other out-of-the-way area to use, and replace it with a close match.

Also, even the toughest carpet out there won't be able to defend itself against a dropped cigarette or sharp knife. Burns and tears are best removed and patched as well.

1 **Prepare area** Since conventional carpet was stretched and put under tension when it was installed, you can't simply cut into it without preparing the area first. What you need to do is release the tension in the area to be patched so that you can safely cut into the carpet without causing a rip or tear. To do this, use a knee kicker to push the carpet toward the patch area, and tack a strip of old carpet to the floor as shown. Work your way around the patch area, relieving tension with the knee kicker and installing the strips.

2 **Cut out damage** With the tension relieved, the next step is to cut out the damaged area. Use a sharp carpet knife for this, and try to keep the cuts as straight as possible. Use the piece you removed as a template to cut an exact-sized replacement patch from a piece of spare carpet (you knew someday you'd have a use for that remnant you saved). Trim any loose pile off the edges of the patch and repair area, and apply seam sealer to both the carpet and the patch to prevent the edges from unraveling.

3 **Insert seam tape** The best way to fasten a patch to the floor or carpet is with special seam-repair tape and adhesive from a flooring supplier or contractor. Cut the tape to fit the cutout, and cover half of each strip with seam adhesive. Then slip each strip under the original carpet's edges. Apply more adhesive to the exposed area, and seat the patch (*see Step 4*). If the patch area is small, you may be able to get away with using professional-grade carpet tape; make sure it is cloth and is at least 2" wide. Apply this to the carpet as you would seam tape.

4 **Seat the patch** Align the patch to fit the cutout, and set it in place. Check the position carefully before pressing down. Press firmly around the edges of the carpet and the patch. Check the drying time of the seam adhesive you used, to make sure you allow sufficient time for the adhesive to dry before removing the tacks and carpet strips. A little extra dry time is a good idea, to make sure the patch is secure before stressing the seams with tension.

SMALL AREAS: TUFT SETTER

Small stains can be repaired with a nifty tool called a tuft setter (*shown*) that's used to replace the damaged fibers with new ones. Flooring suppliers and contractors will special-order one for you if they don't stock them. Before you can use a tuft setter, you'll need to prepare the area by first cutting away the damaged pile with scissors and then swabbing the bare area with latex cement. Remove fibers from a scrap of matching carpet, and fold a fiber into the V of the tool and punch it into the bare area with a few taps of a hammer. Set the new pile so it protrudes slightly, and then use scissors to trim it flush when the repair is complete.

Patching Hardwood Flooring

Hardwood flooring is one of the most durable floor surfaces you can install. But over time, strips can warp or twist and the surface can be damaged deep enough to warrant replacement. There are two main methods used to do this—one simple, the other complex. With the simple method (*as shown below*), you cut a rectangular hole and fill it in with new flooring. The complex method (*see page 113*) involves cutting the hole as a staggered pattern to better conceal the patch. This method requires considerably more work and finesse with a chisel.

   No matter which method you choose, the biggest challenge may be finding replacement flooring to match the existing floor. Here again, pack rats will be delighted that they saved those extra boards. Flooring suppliers may be able to help you find matching flooring. You can also pull up some original flooring to make a patch from a closet or other seldom-seen area, and fill that spot in with a close match.

1 **Plunge cut** The first thing to do to replace a section of hardwood strip flooring is to remove the damaged area. The simplest way I've found to do this is to first make a series of plunge cuts with a circular saw or a trim saw. Set the depth of the blade just slightly deeper than the thickness of the strip flooring, and make three plunge cuts: two to define the ends of the patch, and one down the center to make it easier to remove the damaged area. Take your time with the end cuts to make them as straight as possible.

2 **Pry out middle section** Once you've made the three plunge cuts, remove the middle section. To do this, insert the end of a pry bar into the saw kerf you made in the middle piece. A few taps on the end of the pry bar will drive it far enough under one of the sections that you can pry it out. Repeat for the other half of the middle section. Note: This middle piece may or may not just simply pop out; it'll depend on whether it is nailed to the subfloor.

3 **Finish end cuts** Because the blade of a circular saw is round, it leaves a curved kerf when you make a stopped cut. This means there's still wood that needs to be removed at the ends of the remaining strips. To free these pieces, insert a sharp chisel into the saw kerf near the end and strike it with a hammer. If necessary, split the piece lengthwise with the chisel for better access. Once you've freed the strip, go back with the chisel and pare the edges so they're flat and straight.

4 **Insert new end strips** Now you can insert new end strips. Measure each section separately, and cut pieces of scrap flooring to fit. What you're looking for here is a friction-tight fit. It should be necessary to persuade the piece in with a few gentle taps with a mallet or hammer. Use a short scrap of flooring to prevent damaging the tongue or groove as you slide the piece into position. Tip: Carpenters often cut a slight downward bevel on the ends to allow the piece to slide in easier.

5 **Blind-nail** With one piece in place, drill angled pilot holes for nails into the edge of the strip. Pick a drill bit that's slightly smaller than the diameter of the nail. Nail the strip to the subfloor with 1½" finish nails near the ends of the strip and every 3" to 4". Then countersink the nail heads below the surface with a nail set. Repeat Steps 4 and 5 for the remaining end piece.

6 **Prepare last strip** With both end pieces in place, measure and cut the last strip to length. Before it can be installed, there's one more thing to do. Since you can't slide this last piece in place, you'll need to remove the lower lip of the groove so that you can install it. Place a protective scrap of wood on the floor and lay the last strip on it, face down. Use a sharp chisel to split off the lower lip. Then carefully pare away any sections of the lip that didn't split off with the chisel. Removing the lip like this effectively converts the groove in the strip into a rabbet.

7 **Install last strip** Here's when you'll find out how carefully you measured and cut this final strip. Tilt the middle piece at an angle, and hook the grooved edge of the strip over the tongue in the appropriate end piece. Tap the rabbeted edge with a soft-faced mallet until the groove seats into the tongue. As you do this, you'll need to lower the strip until it's flat. Use the soft-faced mallet or a hammer and a scrap of wood to drive the last strip in place. If there are gaps at the ends, pry the piece out and cut another.

8 **Face-nail** With the patch in place, you can face-nail the last strip to the subfloor. It's always best to drill pilot holes in hardwood—especially near the ends—to prevent splitting. Nail the strip in place using 1½" finish nails, and set them below the surface with a nail set. Sand the new strip smooth, and apply at least two coats of polyurethane. Tip: As you sand, allow the sandpaper to overlap slightly onto the finished flooring. This "feathers" the original finish and will make the patch less noticeable once the finish is applied.

STAGGERED PATTERN

1 Drill holes To remove damaged flooring in a staggered pattern, start by identifying the joints nearest the damage. Then, using a large spade bit, Forsnter bit, or hole saw, drill a series of holes on the waste side of these joints as shown. Mark the drilling depth with a strip of masking tape to match the thickness of the strip flooring so that you won't drill into the subfloor and weaken it.

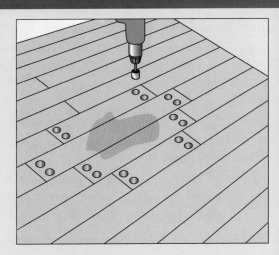

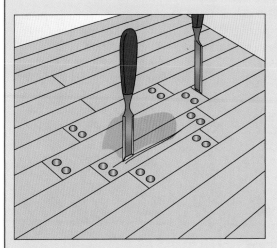

2 Split strips with chisel With a sharp chisel, break the nibs between the holes you drilled to free the ends of the strips. Then split the damaged strips in half lengthwise with a chisel and a mallet or hammer by driving the chisel into the middle of the strip, working your way down along its length. Don't be too aggressive here, as it's easy to damage the adjacent strips, whose surface may crack or split under pressure.

3 Replace pieces Now you can remove the split pieces with a pry bar. Insert the end into a crack in the middle of a board and pry. If necessary, slip a small block of wood under the pry bar for added leverage and to protect adjacent strips. Then trim the edges of the cut area so it's square. Install a patch as you would for a square patch shown on pages 110–112. Here again, you'll need to insert the outermost strips first and then cut the lip off the bottom edge of the groove in the final strip in order to install it.

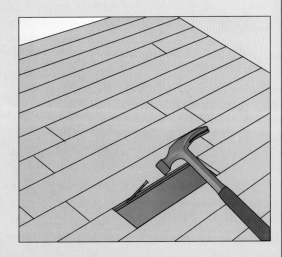

Refinishing a Hardwood Floor

A hardwood floor doesn't always need to be patched to remove scratches and stains—sometimes all it takes is refinishing. If a scratch or stain is small, you may be able to remove it by scraping and sanding (*see page 116*). But if the entire floor is worn out, give it a facelift with a rented floor sander. There's nothing complicated about refinishing a wood floor. The biggest hassle is dealing with the clouds of dust you'll generate.

One way you can reduce this is to use an orbital floor sander instead of a drum-style floor sander. The more-aggressive drum sander will do the job quicker, but at a price. Drum sanders generate copious amounts of dust, and it's easy to accidentally sand a groove or dip into your floor. That's why I suggest an orbital sander—it's slow, not as aggressive, and virtually foolproof. Even if you use an orbital sander, you'll still generate a lot of dust. Hold this to a minimum with proper room preparation (*see below*).

1 **Room preparation** To prevent dust from covering everything in your home, take the time to prepare the room to be refinished. Start by removing the furniture, window treatments, and anything on the walls. If there's an overhead light, remove the diffuser. Cover electrical outlets and switches with masking tape to keep them free from dust. Then attach a plastic drop cloth from ceiling to floor with masking tape. Overlap two drop cloths or cut a slit in the plastic to serve as a door. Attach a furnace filter to the intake of a square fan and place it in an open window to pull dust out of the room.

2 **Orbital sander** For refinishing floors, I suggest using sanding screen instead of sandpaper. Even coarse sandpaper will quickly clog up with old finish. But sanding screen, with its open weave, will go a lot longer without clogging. Place a piece of screen beneath the sander's pad, and slip the electrical cord over your shoulder to keep it out of harm's way. Start in one corner and work a 3-foot-square section at a time. Don't let the sander come in contact with the wall or baseboards, as it will likely cause damage—I try to stay about 3" to 4" away. (I sand edges with a portable sander; *see Step 3*.)

3 **Random-orbit sander for details** Once you've sanded the entire floor, use a random-orbit sander to remove the remaining finish around the perimeter of the room. If your sander accepts hook-and-loop sandpaper, you can use sanding screen on it by sandwiching a piece of abrasive pad between the sanding pad and the screen. The Velcro-like hooks of the sanding pad will grab the fibers of the abrasive pad and hold it firmly in place. If the sander uses pressure-sensitive adhesive (PSA) paper, start with a fairly coarse grit (60 to 80) and work up to 120-grit sandpaper.

4 **Vacuum** When you've removed the old finish and you're satisfied with the smoothness of the floor, it's time to vacuum, vacuum, and then vacuum. Yes, I'm serious. The number one cause of a poor finish is that the floor wasn't properly cleaned before the finish was laid down. With a shop-vacuum, go over the entire floor, empty the vacuum, clean the filter (this is important because a dirty filter will blow a lot of dust back into the room), and repeat. Allow the dust to settle for a couple of hours, and then vacuum again.

5 **Apply finish** With the floor as clean as possible, you can apply the finish. Use a polyurethane that's formulated for floors. I like to use a lamb's-wool applicator and apply a minimum of three coats. Check the polyurethane can to see how long you should wait between coats. Keep your applicator pliable between coats by tightly wrapping it with a plastic bag, taped shut so air can't get inside. Allow the final coat to cure at least one full day before moving furniture back into the room. Protect your beautiful new floor by attaching felt pads to furniture legs.

Removing Stains in Hardwood Flooring

1 **Scrape** Small stains or scratches can be removed from a strip hardwood floor with a bit of elbow grease. Start by using a paint scraper or cabinet scraper to remove any surface stains. Use as little pressure as possible, as both these tools have a tendency to dig in at the corners. Keep the blade of the scraper flat on the floor and tilt it at an angle for best results. Stop and check the blade periodically, and remove any gummed-up paint or finish with a clean, soft cloth.

2 **Feather-sand** When you've scraped away as much as possible, use a sanding block and some 100-grit sandpaper to sand the floor until the stain or scratch disappears. Then "feather" the edges surrounding the old stain or scratch by allowing the sandpaper to overlap onto the finished area. Sand gently here to make a smooth transition between the old finish and the bare wood. Feathering like this makes the newly finished area less noticeable and roughs up the old finish so it'll bond better with the fresh finish.

3 **Apply finish** After you've smoothed the edges of the repair area, it's time to apply a couple coats of finish. Brush on a thin first coat (a foam brush works great for this) and let it dry. Apply at least one more coat, but better yet three, allowing the finish to dry between coats. Check the can for the manufacturer's recommended drying time. Tip: Professional floor refinishers will sand lightly in between coats with 220-grit wet/dry paper. This levels the surface and removes any impurities the wet finish may have attracted as it dried.

Repairing Parquet Tiles

1 **Drill hole** Small stains or burns (like the one shown here) in parquet tiles can be repaired without replacing the entire tile. Instead, the individual damaged sections can be easily removed and replaced with a patch. To do this, start by selecting a drill bit that's slightly smaller than the thickness of one strip (or fillet). Define the ends of the patch by drilling through the tile as shown (a Forstner bit works great for this). Don't drill into the subfloor, or you'll weaken it.

2 **Remove damaged piece** Next, pry out the damaged piece with a flat-blade screwdriver. Then use a sharp chisel to square up the ends of the hole. With an old screwdriver or small putty knife, scrape up any old adhesive from the subfloor. Measure the hole and cut a piece to fit. If you've got some extra tiles lying around, try to select a patch piece that will blend in well. Spread floor adhesive on the subfloor, and persuade the patch piece in with a soft-faced mallet (*inset*). You're after a nice snug fit. If it's loose, pry it out and cut another patch.

REPLACING A FULL TILE: CUT AND REMOVE

If you need to remove a full parquet tile, set the blade depth on your circular saw to the thickness of the tile. Then make a series of plunge cuts about 1" from the edge of the tile. Take care not to cut into an adjacent tile. Slip the business end of pry bar into one of the saw kerfs, and pry up. Stubborn tiles may require a chisel and a hammer to persuade them to leave. Scrape the subfloor clean and apply fresh adhesive. As you did with strip flooring (*see page 112*), you'll need to chisel off the lower lip of the groove in order to set the new tile in place.

Repairing Sheet Flooring

Stains, tears, and cracks in resilient sheet flooring can be repaired with a little patience and some suitable replacement flooring. The most reliable way to do this is to cut out the damage and replace it with a patch, *as shown below.* **A word of caution here:** Resilient flooring and the adhesive used to attach it manufactured prior to 1986 may contain asbestos. If in doubt, have a sample piece of the flooring tested by an independent laboratory. Contact your local flooring contractor or supplier to find the nearest lab. If you don't want to go to this trouble, then assume that it is asbestos flooring and take the appropriate precautions.

Wear a dual-cartridge respirator at all times. Asbestos fibers will go right through an ordinary dust mask. Never sand the flooring or underlying adhesive. If you must remove it, keep the surface wet to reduce airborne dust. Contact your local environmental protection agency for guidelines for disposal.

1 **Tape patch over damage** To create a patch to repair resilient flooring, start by measuring the damaged area and cut a replacement piece large enough to cover it. Make sure to carefully match the pattern if there is one, and then tape the patch over the damaged area. Take your time here, as this step will determine how "invisible" the patch will be. As you tape the replacement piece in place, press it down firmly against the flooring to remove any air pockets.

2 **Cut through both layers** After you've got the replacement piece in place, use a sharp utility knife to cut through both the new piece and the old flooring at the same time. Whenever possible, make the cut along a "grout" line—try to split the grout line in half as you cut. Cutting both pieces simultaneously ensures that the patch will be a perfect fit in the hole.

3 **Remove old flooring** Remove the tape and set the excess piece aside. Before you either move the patch or remove the damaged area below, place a piece of tape on one edge of the patch and the corresponding edge of the hole as a reference for alignment. Then set the patch aside and, if the tile was installed with adhesive, use a wide-blade putty knife to remove the damaged area; scrape the subfloor to remove any old adhesive. If the tile is a perimeter-install, you'll be able to lift the patch right out.

4 **Spread adhesive and install patch** Using the appropriate-sized notched trowel, spread flooring adhesive onto the subfloor. Avoid applying any within ½" of the edges, to help reduce squeeze-out. Match up the reference tape on the patch with the tape you applied to the floor, and press the patch in place. Use a small hand roller or rolling pin to press the tile firmly into the adhesive. Wipe up any excess with a clean cloth dampened with solvent. Place a weight on the patch overnight to help it lie flat. Then seal the seams with seam sealer, or use the trick shown below.

BONDING THE EDGES OF A PATCH

Here's a nifty trick that if done properly will make your newly installed patch nearly invisible. Cover the edges of the patch with a double layer of aluminum foil with the dull side down. Then turn an iron on to high and allow it to warm up. Press the iron on the foil over the seam several times. Check often to make sure you're not discoloring the tile. If done carefully, the iron will partially melt the cut edges of the seam, allowing them to join together to form a solid bond.

Repairing Vinyl Tiles

One of the advantages a vinyl tile floor offers is that you can replace tiles that get damaged without worrying about hiding a patch. By its very nature, a vinyl tile floor has many seams. As long as you replace a full tile at a time, there will be no additional seams to catch the eye.

The high number of seams, however, does increase the likelihood of problems—particularly loose tiles and broken corners. Fortunately, loose tiles are easy to replace (*see below*).

That's not to say repairng vinyl tile is without its challenges. As with all repairs, matching the tile can be a problem. If you've got tiles left over from the original installation, you're in good shape. Otherwise, you'll have to visit flooring suppliers to find a close match. Also, old tile that's been installed for a long time can be very stubborn to remove. See the tips on page 121 for ways to deal with this. Finally, any tile floor laid before 1986 may contain asbestos; see page 118 for more on how to check for this and ways to handle it.

1 **Secure loose tiles** As long as a loose tile is in good shape, you may be able to refasten it. If the tile is old and brittle, warm it with a heat gun to make it flexible. This way you'll be able to bend it back to do two things. First, if the tile has been loose for a while, odds are that dirt has accumulated under it. To get a solid bond, this dirt (along with the old adhesive) needs to be removed. Second, it allows you to better spread fresh adhesive under the tile. Press the reglued tile in place, and clean up any excess glue immediately. Set a weight on the tile overnight to keep it flat.

2 **Deflating an air pocket** Air pockets are common in vinyl tiles. If left unattended, the protruding portion of the tile will harden and crack and split when stepped on. To prevent this from happening, take action as soon as you discover a bubble. Cut a slit in the top of the bubble with a sharp utility knife to allow the air to escape. Then, depending on the size of the bubble, force in some fresh adhesive with a toothpick, a screwdriver tip, or a putty knife. Wipe up any excess adhesive, and place a weight on the bubble overnight.

3 **Heat to remove** If a tile is damaged to a point where it needs to be replaced, you'll need to persuade the old adhesive to give up its grip on the tile. There are a couple ways to do this—both ways use heat to soften the old adhesive. The method I've found that works best is to warm the tile with a heat gun, working from the edges in. Insert a putty knife under a loose edge and pry up the tile. Another option is to use an iron and a piece of aluminum foil to warm the tile as described on page 119. But be careful—it's easy to scorch the surrounding tiles.

4 **Scrape off old adhesive** Once you've pried off the damaged tile, use a stiff-blade putty knife to scrape off the old adhesive from the subfloor. Pay particular attention to the edges of the surrounding tile. In order for the new tile to lie flat and the seams to be tight, the subfloor near the edges needs to be spotless. A razor-edged scraper (such as a wallpaper scraper) does a good job of scraping down to bare wood. When you're done scraping, vacuum the subfloor thoroughly to remove any bits of dust, dirt, and old adhesive.

5 **Install new tile** Now you can insert the new tile. The directional arrow on the back of the tile won't be any help here since you don't know how the other tiles were laid. Look at the pattern on the new tile and the old flooring closely, and try to match the pattern as best you can. Press the new tile in place, and wipe up any excess adhesive with a clean cloth dampened in solvent. Place a weight on the tile overnight to keep it flat until the adhesive has set up.

Removing Tile Stains

Remove mildew To remove mildew stains from white grout, dip a toothbrush in full-strength household chlorine bleach and scrub the area clean. (Wear old clothes when you do this, as a toothbrush will dispense a fine spray of bleach in all directions as you scrub.) To get rid of the chlorine smell when you're done, wipe on a solution of baking soda and water. Let the grout dry completely (two days), and then apply sealer to help prevent mildew from quickly growing back. Note: don't apply bleach to colored grout; use a commercial cleaner designed for this purpose.

Heavy-duty cleaners For really stubborn stains, use a commercially available cleaner. Wear rubber gloves, eye protection, and a cartridge-style mask. Read and follow the instructions on the label. Many of these heavy-duty cleaners are acid-based and are dangerous to work with. Make sure there's adequate ventilation when you go to apply the cleaner. Start with a weak solution, and increase the strength if that doesn't work. Rinse off the cleaner thoroughly when done.

How to Remove Stains from Tiles

Stain	How to Remove
Blood, coffee, juice, lipstick, rust ,tea, wine	Mix baking powder with water to form a thick paste. Spread the paste on the stain and let it dry. Rinse the area and wipe dry. If the stain persists, apply full-strength chlorine bleach.
Cooking fats or grease	Use a sponge and a concentrated solution of all-purpose household cleaner to scrub the stained area.
Mineral deposits	Apply a solution of half ammonia and half water, or full-strength white vinegar. Rinse and pat dry.
Paint	Wipe a commercial paint remover on the stain (test an out-of-the-way tile first to make sure the remover doesn't harm the tile). Once the paint has softened, scrape it off with a razor knife.

Replacing Grout

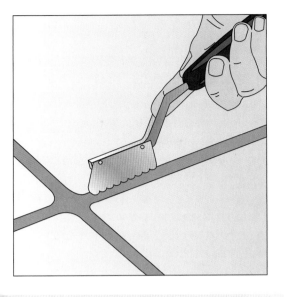

1 **Remove old grout** The best way I've found to remove old grout is to use a grout saw (*see the drawing*). Small bits of tungsten carbide are embedded into the business end of the saw so that it can easily grind through the grout. The blades on most grout saws are replaceable. To remove thick grout lines, you either stack up standard blades or purchase a wide blade designed specifically for this. In a pinch, you can remove grout with a utility knife or an old hacksaw blade.

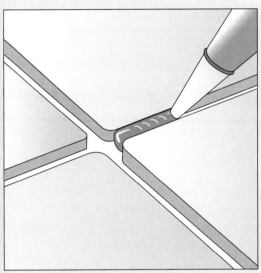

2 **Apply new grout** Vacuum the joints and clean the surrounding tiles. Then mix up a small amount of matching grout and fill a grout bag. If you're repairing only a small section, press the grout in place with a plastic putty knife or your finger. If you're repairing a large section, use a grout float held at an angle to skim off the excess grout (*see page 56 for more on this*). Allow the grout to dry to a haze.

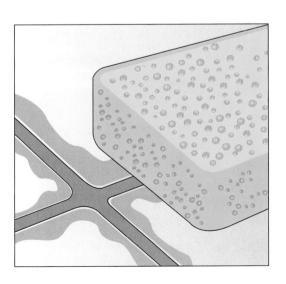

3 **Clean and seal** Remove the haze of the dried grout from the tiles with a damp sponge, using a circular motion. Take care not to work a grout line excessively; if you do, you're likely to pull the grout right out of the joint. Try to match the depth of the grout joints to the existing grout. After the grout has completely dried, apply grout sealer, taking care to apply it just to the grout. Wipe off any excess immediately with a clean, soft cloth.

Replacing Ceramic Tiles

If grout is crumbling and tiles are loose, the first thing you should do is try to find out what's causing the problem. In many cases, moisture is the culprit. Check the floor carefully (from underneath, whenever possible) to find the source of the moisture. Call in a flooring contractor if necessary. Replacing tile is fairly straightforward work. If tiles are loose, half the job is done. If, on the other hand, tiles are cracked or damaged but still intact, removing them will likely be your biggest challenge.

The single most important piece of advice I can offer concerning removing well-fastened ceramic tile is to wear eye protection. Standard eyeglasses won't do. You need approved wrap-around glasses or, even better, goggles. The tiniest ceramic chip in your eye can cause a lot of pain and even serious eye damage. Protect your eyes—they're the only pair you have.

1 **Break out old tile** Wearing safety glasses, use a cold chisel and a hammer to break out the damaged tile (small chips of ceramic tile will most certainly fly off in all directions). Start by tapping the tile with the hammer to crack it into smaller pieces. Then insert the chisel into one of the cracks and tap gently. Take care not to use excessive force: It's easy to damage the substrate below and loosen other tiles. Work the chisel from the center of the tile out toward the edges.

2 **Scrape the surface** Once you've removed the majority of the old tile, use a stiff-blade putty knife to scrape away any remaining tile and the old adhesive. Use it also to chip away and remove the grout from the joints. Take extra care if the substrate beneath the tile is cement board. It's easy to get carried away with scraping and end up scraping away some of the cement board. If you hit the mesh that holds the cement board together, you've gone too far. Fill any depressions with thin-set mortar and allow it to dry before applying adhesive.

3 **Apply adhesive** Mix up enough thin-set mortar to attach the tile or tiles. For small repairs like the one shown, you can use flooring adhesive. Apply the adhesive or thin-set mortar with the appropriate-sized notched trowel. You can either apply the adhesive to the subfloor (*as shown here*) or spread it directly onto the back of the tiles—a technique referred to as back-buttering. If you're using adhesive, let it set for 15 minutes to get tacky before installing the tile.

4 **Set the tile** Position the tile, matching up any relevant patterns, and drop it onto the adhesive or thin-set mortar. Press the tile firmly in place and twist it slightly to help spread the adhesive. Then gently tap the tile with a soft-faced mallet to set or bed the tile into the adhesive. Optionally, use a padded scrap of wood—a "beater" block—and a hammer to set the tile flush with the surrounding tiles. Wait the suggested number of days for the adhesive or mortar to cure before applying grout.

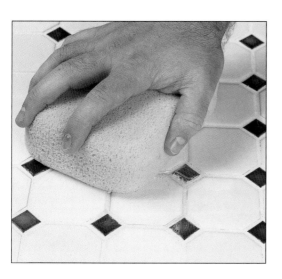

5 **Grout and seal** Mix up and apply matching grout to the joints (*see page 123 for more on this*). Allow the grout to haze over, and then wipe away the haze with a damp sponge. Match the depth of the new grout lines to that of the existing grout. After the grout has dried completely (check the label on the grout for suggested drying times), brush sealer onto the grout. Wipe up any excess immediately with a clean, soft cloth.

Glossary

Back-buttering – a controlled technique for applying adhesive or thin-set mortar to the back of a tile, much like buttering toast, instead of applying it to the floor or wall.

Backer board – a cement-based panel that's reinforced with fiberglass mesh and used as an underlayment for tile.

Battens – thin wood strips temporarily attached to the subfloor to serve as a reference to ensure straight courses of ceramic tile, parquet tile, or strip flooring, held in place with adhesive.

Beater block – a scrap of wood that's covered with carpet or other protective material and used with a hammer to level tiles.

Blind-nailing – a nailing technique where the head of the nail is concealed from view; used to attach hardwood strip flooring to the subfloor by nailing through the tongue, which is concealed by the next strip.

Border tiles – partial tiles that run along the perimeter of the room and fill the space between the full tiles and the wall.

Bridging – cross-brace supports, either metal or wood, that fit between joists to reinforce the floor.

Carpet – any of a variety of flooring where loops of yarn are forced through a fiber backing; the loops may be left in place or trimmed. Conventional carpet is stretched across the room and held in place with tackless strips; cushion-backed carpet is glued in place with flooring adhesive.

Carpet padding – a layer of foam that's installed between conventional carpet and the subfloor to cushion one's step.

Carpet stretcher – a specialized tool, usually rented, that's used to stretch conventional carpet taut before it's attached to tackless strips around the room's perimeter.

Casing – boards that line the inside of a doorway, or the trim around the door opening.

Cement board – *see Backer board*

Ceramic tile – any tile made from refined clay that's mixed with additives and water and hardened in a kiln. Ceramic tile can be either glazed or unglazed.

Contour gauge – a special layout tool used to transfer profiles to flooring. It consists of a series of friction-tight "fingers" that conform to a profile when pressed against the obstacle, such as a door casing.

Countersinking – a technique used to set a fastener below the work surface, using either a countersink bit in a drill or a nail set with a hammer.

Cove base molding – preformed vinyl trim molding with a cove at the bottom to serve as a transition between a wall and flooring; available in either strips or rolls.

Cushion-backed carpet – a type of carpet where a layer of foam is attached to the back of the carpet, instead of using a separate pad. This type of carpet is glued to the floor with flooring adhesive.

Face-nailing – a nailing technique which leaves the nails in view unless countersunk and filled. Typically used to fasten the first strip of hardwood flooring against a wall and to install border trim.

Felt paper – also referred to as building paper, used as a cushion or vapor barrier between the subfloor and flooring, typically hardwood strip flooring.

"Floating" floor – *see Laminate flooring*

Flooring nailer – a special fastening tool used to install hardwood strip flooring. It shoots a barbed fastener into the tongue of the flooring to secure it to the subfloor.

Floor roller – a heavy roller (typically 75 or 100 pounds) that can be rented to press full-adhesive flooring firmly onto the adhesive to ensure a solid bond.

Floor sander – a rental tool used to sand or refinish hardwood flooring. A drum sander is fast but aggressive; an orbital sander is slower but easier to use.

Full-adhesive installation – any installation where adhesive is spread over the entire subfloor prior to laying the flooring. The flooring is then pressed into the adhesive with a floor roller.

Grout – a thin mortar mixture used to fill the gap between ceramic tiles.

Grout bag – a special, reusable bag with a metal tip that's filled with grout and then squeezed to apply a bead of grout to the gaps between tiles.

Grout float – a rubber-backed trowel used to force grout into the gaps between tiles, and then used to skim off any excess.

Grout saw – a special saw with bits of tungsten carbide embedded in the blade, used to grind away old grout when doing repairs.

Joist – a horizontal structural member used to support either a floor or a ceiling.

Knee kicker – a rental tool used to stretch conventional carpet near walls and force the carpet to lie flat in the corners.

Laminate flooring – a man-made floor panel that's a sandwich of a fiberboard core laminated with melamine to imitate other flooring. The panel edges are glued together, and the room-sized panel "floats" on the subfloor—it's not attached at all.

Mosaic tile – ceramic tile, usually 2" or smaller, that's sold in sheets of tiles held together with a mesh backing.

Parquet tile – wood flooring made in squares with a variety of patterns. The edges are tongue-and-grooved for ease of installation.

Perimeter installation – an installation method used for resilient sheet flooring, where the flooring is attached to the subfloor only around the perimeter.

Pile – the tufts of yarn, either cut or looped, that form the surface of a carpet.

Reducer strip – a strip of wood with a curved edge that's used as a transition between flooring of differing thicknesses.

Resilient flooring – flooring made of vinyl, vinyl composition, or rubber. Can be purchased in either sheet form or squares.

Ring-shank nails – nails that are manufactured with rings along the shank to provide extra grip.

Scribing – a layout technique used to copy the imperfections of a wall to flooring so the flooring can be cut to butt snugly against the wall.

Seam tape – a wide strip of tape covered with hot-melt adhesive and used to join pieces of carpet. A seaming iron (usually rented) is used to melt the glue.

Self-adhesive tile – any type of flooring squares that have an adhesive applied to the back, covered by a paper backing. Also referred to as peel-and-stick.

Shim – a thin piece of wood. When driven behind a surface, it forces the surface to become level or plumb.

Shoe molding – wood strip molding that's commonly used to conceal an expansion gap between base molding and the floor.

Strap clamps – special clamps used to install laminate flooring. An adjustable strap provides the clamping pressure when it's tightened with the built-in ratchet.

Subfloor – the first layer of a floor structure, fastened directly to the joists or to a concrete slab.

Substrate – the surface or underlayment to which tile is applied; typically cement board.

"Tackless" strips – thin strips of plywood with protruding tacks that are installed around the perimeter of a room to hold conventional carpet in place.

Thin-set mortar – a cement-based adhesive used to attach tile to cement board or other underlayment, or to attach cement board to the subfloor.

Tile cutter – a special tool that uses a wheel to score a line on a tile; the tile is then snapped in two at the scored line.

Tile nipper – a plier-type tool that is used to nip or break off small pieces of tile.

Tile saw – can be motorized or manual. Manual tile saws are often referred to as rod saws.

Transition strip – a strip of metal or wood that serves to delineate two types of flooring, usually at a threshold of a door.

Trowel – any of several flat and oblong tools that are used for handling grout, adhesive, or mortar. It may have smooth or notched edges.

Tongue-and-groove flooring – strips of wood flooring that have a protruding tongue on one edge and a matching groove on the other for the purpose of joining several strips together.

Underlayment – a smooth surface laid on top of the subfloor to accept flooring; can be sheets of plywood, foam, cork, or cement board.

Vapor barrier – plastic or other sheeting installed between layers of flooring to prevent moisture from damaging the flooring.

Index